Thermal and Statistical Physics Simulations
The Consortium for Upper-Level Physics Software

Harvey Gould
Department of Physics, Clark University,
Worcester, Massachusetts

Lynna Spornick
Applied Physics Laboratory, Johns Hopkins University,
Baltimore, Maryland

Jan Tobochnik
Department of Physics, Kalamazoo College,
Kalamazoo, Michigan

Series Editors

Robert Ehrlich
Maria Dworzecka
William MacDonald

JOHN WILEY & SONS, INC.

NEW YORK · CHICHESTER · BRISBANE · TORONTO · SINGAPORE

ACQUISITIONS EDITOR Cliff Mills
MARKETING MANAGER Catherine Faduska
SENIOR PRODUCTION EDITOR Sandra Russell
MANUFACTURING MANAGER Mark Cirillo

This book was set in 10/12 Times Roman by Beacon Graphics and
printed and bound by Hamilton Printing Co. The cover was printed by Phoenix Color.

Recognizing the importance of preserving what has been written, it is a
policy of John Wiley & Sons, Inc. to have books of enduring value published
in the United States printed on acid-free paper, and we exert our best
efforts to that end.

The paper on this book was manufactured by a mill whose forest management programs include
sustained yield harvesting of its timberlands. Sustained yield harvesting principles ensure that
the number of trees cut each year does not exceed the amount of new growth.

Library of Congress Cataloging in Publication Data:
Gould, Harvey
Thermal and statistical simulations : the consortium for upper level
physics software / Harvey Gould, Lynna Spornick, Jan Tobochnik
Includes bibliographical references (p.).
ISBN 0-471-54886-3 (pbk./disk)

Printed in the United States of America

10 9 8 7 6 5 4 3 2

Contents

List of Figures

List of Tables

1

Introduction

"It is nice to know that the computer understands the problem. But I would like to understand it too."

—Eugene P. Wigner, quoted in *Physics Today,* July 1993

1.1 Using the Book and Software

The simulations in this book aim to exploit the capabilities of personal computers and provide instructors and students with valuable new opportunities to teach and learn physics, and help develop that all-important, if somewhat elusive, physical intuition. This book and the accompanying diskettes are intended to be used as supplementary materials for a junior- or senior-level course. Although you may find that you can run the programs without reading the text, the book is helpful for understanding the underlying physics, and provides numerous suggestions on ways to use the programs. *If you want a quick guided tour through the programs, consult the "Walk Throughs" in Appendix A.* The individual chapters and computer programs cover mainstream topics found in most textbooks. However, because the book is intended to be a supplementary text, no attempt has been made to cover all the topics one might encounter in a primary text.

Because of the book's organization, students or instructors may wish to deal with different chapters as they come up in the course, rather than reading the chapters in the order presented. One price of making the chapters semi-independent of one another is that they may not be entirely consistent in notation or tightly cross-referenced. Use of the book may vary according to the taste of the student or instructor. Students may use this material as the basis of a self-study course. Some instructors may make homework assignments from the large number of exercises in each chapter or to use them as the basis of student projects. Other instructors may use the computer programs primarily for in-class demonstrations. In this latter case, you may find that the programs are suitable for a range of courses from the introductory to the graduate level.

Use of the book and software may also vary with the degree of computer programming performed by users. For those without programming experience, all the computer simulations have been supplied in executable form, permitting them to be used as is. On the other hand, Pascal source code for the programs has also been provided, and a number of exercises suggest specific ways the programs can be modified. Possible modifications range from altering a single procedure especially set up for this purpose by the author, to larger modifications following given examples, to extensive additions for ambitious projects. However, the intent of the authors is that the simulations will help the student to develop intuition and a deeper understanding of the physics, rather than to develop computational skills.

We use the term "simulations" to refer to the computer programs described in the book. This term is meant to imply that programs include complex, often realistic, calculations of models of various physical systems, and the output is usually presented in the form of graphical (often animated) displays. Many of the simulations can produce numerical output—sometimes in the form of output files that could be analyzed by other programs. The user generally may vary many parameters of the system, and interact with it in other ways, so as to study its behavior in real time. The use of the term simulation should not convey the idea that the programs are bypassing the necessary physics calculations, and simply producing images that look more or less like the real thing.

The programs accompanying this book can be used in a way that complements, rather than displaces, the analytical work in the course. It is our belief that, in general, computational and analytical approaches to physics can be mutually reinforcing. It may require considerable analytical work, for example, to modify the programs, or really to understand the results of a simulation. In fact, one important use of the simulations is to suggest conjectures that may then be verified, modified, or proven false analytically. A complete list of programs is given in Section 1.7.

1.2 Required Hardware and Installation of Programs

The programs described in this book have been written in the Pascal language for MS-DOS platforms. The language is Borland/Turbo Pascal, and the minimum hardware configuration is an IBM-compatible 386-level machine preferably with math coprocessor, mouse, and VGA color monitor. In order to accommodate a wide range of machine speeds, most programs that use animation include the capability to slow down or speed up the program. To install the programs, place disk number 1 in a floppy drive. Change to that drive, and type Install. You need only type in the file name to execute the program. Alternatively, you could type the name of the driver program (the same name as the directory in which the programs reside), and select programs from a menu. A number of programs write to temporary files, so you should check to see if your autoexec.bat file has a line that sets a temporary directory, such as SET TEMP = C:\TEMP. (If you have installed WINDOWS on your PC, you will find that such a command has already been written into your autoexec.bat file.) If no such line is there, you should add one.

Compilation of Programs

If you need to compile the programs, it would be preferable to do so using the Borland 7.0 (or later) compiler. If you use an earlier Turbo compiler you may run out of memory when compiling. If that happens, try compiling after turning off memory resident programs. If your machine has one, be sure to compile with the math-coprocessor turned on (no emulation). Finally, if you recompile programs using any compiler other than Borland 7.0, you will get the message: "EGA/VGA Invalid Driver File" when you try to execute them, because the driver file supplied was produced using this version of the compiler. In this case, search for the file BGILINK.pas included as part of the compiler to find information on how to create the EGAVGA.obj driver file. *If any other instructions are needed for installation, compilation, or running of the programs, they will be given in a README file on the diskettes.*

1.3 User Interface

To start a program, simply type the name of the individual or driver program, and an opening screen will appear. All the programs in this book have a common user interface. Both keyboard and mouse interactions with the computer are possible. Here are some conventions common to all the programs.

Menus: If using the *keyboard*, press **F10** to highlight one of menu boxes, then use the **arrow** keys, **Home**, and **End** to move around. When you press **Return** a submenu will pull down from the currently highlighted menu option. Use the same keys to move around in the submenu, and press **Return** to choose the highlighted submenu entry. Press **Esc** if you want to leave the menu without making any choices.

If using the *mouse* to access the top menu, click on the menu bar to pull down a submenu, and then on the option you want to choose. Click anywhere outside the menus if you want to leave them without making any choice. Throughout this book, the process of choosing submenu entry **Sub** under main menu entry **Main** is referred to by the phrase "choose **Main | Sub**." The detailed structure of the menu will vary from program to program, but all will contain **File** as the first (left-most) entry, and under **File** you will find **About CUPS**, **About Program**, **Configuration**, and **Exit Program**. The first two items when activated by mouse or arrows keys will produce information screens. Selecting **Exit Program** will cause the program to terminate, and choosing **Configuration** will present you with a list of choices (described later), concerning the mode of running the program. In addition to these four items under the **File** menu, some programs may have additional items, such as **Open**, used to open a file for input, and **Save**, used to save an output file. If **Open** is present and is chosen, you will be presented with a scrollable list of files in the current directory from which to choose.

Hot Keys: Hot keys, usually listed on a bar at the bottom of the screen, can be activated by pressing the indicated key or by clicking on the hot key bar with the mouse. The hot key **F1** is reserved for help, the hot key **F10** activates the menu bar. Other hot keys may be available.

Sliders (scroll-bars): If using the *keyboard*, press **arrow** keys for slow scrolling of the slider, **PgUp/PgDn** for fast scrolling, and **End/Home** for moving from one end to another. If you have more then one slider on the screen then only the slider with marked "thumb" (sliding part) will respond to the above keys. You can toggle the mark between your sliders by pressing the **Tab** key.

If using the *mouse* to adjust a slider, click on the thumb of the slider, drag it to desired value, and release. Click on the arrow on either end of the slider for slow scrolling, or in the area on either side of thumb for fast scrolling in this direction. Also, you can click on the box where the value of the slider is displayed, and simply type in the desired number.

Input Screens: All input screens have a set of "default" values entered for all parameters, so that you can, if you wish, run the program by using these original values. Input screens may include circular radio buttons and square check boxes, both of which can take on Boolean, i.e., "on" or "off," values. Normally, check boxes are used when only one can be chosen, and radio buttons when any number can be chosen.

If using the *keyboard*, press **Return** to accept the screen, or **Esc** to cancel it and lose the changes you may have made. To make changes on the input screen by keyboard, use **arrow** keys, **PgUp**, **PgDn**, **End**, **Home**, **Tab**, and **Shift-Tab** to choose the field you want to change, and use the backspace or delete keys to delete numbers. For Boolean fields, i.e., those that may assume one of two values, use any key except those listed above to change its value to the opposite value.

If you use the *mouse*, click [OK] to accept the screen or [Cancel] to cancel the screen and lose the changes. Use the mouse to choose the field you want to change. Clicking on the Boolean field automatically changes its value to the opposite value.

Parser: Many programs allow the user to enter expressions of one or more variables that are evaluated by the program. The function parser can recognize the following functions: absolute value (abs), exponential (exp), integer or fractional part of a real number (int or frac), real or imaginary part of a complex number (re or im), square or square root of a number (sqr or sqrt), logarithms—base 10 or e (log or ln)—unit step function (h), and the sign of a real number (sgn). It can also recognize the following trigonometric functions: sin, cos, tan, cot, sec, csc, and the inverse functions: arcsin, arccos, arctan, as well as the hyperbolic functions denoted by adding an "h" at the end of all the preceding functions. In addition, the parser can recognize the constants $pi, e, i(\sqrt{-1})$, and rand (a random number between 0 and 1). The operations +, −, *, /, ^ (exponentiation), and !(factorial) can all be used, and the variables r and c are interpreted as $r = \sqrt{x^2 + y^2}$ and $c = x + iy$. Expressions involving these functions, variables, and constants can be nested to an arbitrary level using parentheses and brackets. For example, suppose you entered the following expression: **h(abs(sin(10*pi* x))−0.5)**. The parser would interpret this function as $h(|sin(10\pi x)| - 0.5)$. If the program evaluates this function for a range of x-values, the result, in this case, would be a series of square pulses of width 1/15, and center-to-center separation 1/10.

Help: Most programs have context-sensitive help available by pressing the **F1** hot key (or clicking the mouse in the **F1** hot key bar). In some programs help is available by choosing appropriate items on the menu, and in still other programs tutorials on various aspects of the program are available.

1.4 The CUPS Project and CUPS Utilities

The authors of this book have developed their programs and text as part of the Consortium for Upper-Level Physics Software (CUPS). Under the direction of the three editors of this book, CUPS is developing computer simulations and associated texts for nine junior- or senior-level courses, which comprise most of the undergraduate physics major curriculum during those two years. A list of the nine CUPS courses, and the authors associated with each course, follows this section. This international group of 27 physicists includes individuals with extensive backgrounds in research, teaching, and development of instructional software.

The fact that each chapter of the book has been written by a different author means that the chapters will reflect that individual's style and philosophy. Every attempt has been made by the editors to enhance the similarity of chapters, and to provide a similar user interface in each of the associated computer simulations. Consequently, you will find that the programs described in this and other CUPS books have a common look and feel. This degree of similarity was made possible by producing the software in a large group that shared a common philosophy and commitment to excellence.

Another crucial factor in developing a degree of similarity between all CUPS programs is the use of a common set of utilities. These CUPS utilities were written by Jaroslaw Tuszynski and William MacDonald, the former having responsibility for the graphics units, and the latter for the numerical procedures and functions. The numerical algorithms are of high quality and precision, as required for reliable results. CUPS utilities were originally based on the M.U.P.P.E.T. utilities of Jack Wilson and E.F. Redish, which provided a framework for a much expanded and enhanced mathematical and graphics library. The CUPS utilities (whose source code is included with the simulations with this book), include additional object-oriented programs for a complete graphical user interface, including pull-down menus, sliders, buttons, hot-keys, and mouse clicking and dragging. They also include routines for creating contour, two-dimensional (2-D) and 3-D plots, and a function parser. The CUPS utilities have been provided in source code form to enable users to run the simulations under future generations of Borland/Turbo Pascal. If you do run under future generations of Turbo or Borland Pascal on the PC, the utilities and programs will need to be recompiled. You will also need to create a new egavga.obj file which gets combined with the programs when an executable version is created—thereby avoiding the need to have separate (egavga.bgi) driver files. These CUPS utilities are also available to users who wish to use them for their own projects.

One element not included in the utilities is a procedure for creating hard copy based on screen images. When hard copy is desired, those PC users with the appropriate graphics driver (graphics.com), may be able to produce high-quality screen images by depressing the **PrintScreen** key. If you do not have the graphics software installed to get screen dumps, select **Configuration | Print Screen**, and follow the directions. Moreover, public domain software also exists for capturing screen images, and for producing PostScript files, but the user should be aware that such files are often quite large, sometimes over 1 MB, and they require a PostScript printer driver to produce.

One feature of the CUPS utilities that can improve the quality of hard copy produced from screen captures is a procedure for switching colors. This capability is important because the grey scale rendering of colors on black-and-white printers may create poor contrasts if the original (default) color assignments are used. To access the CUPS utility for changing colors, the user need only choose **Configuration** under the **File** menu when the program is first initiated, or at any later time. Once you have chosen **Configuration**, to change colors you need to click the mouse on the **Change Colors** bar, and you will be presented with a 16 by 16 matrix of radio buttons that will allow you to change any color to any other color, or else to use predefined color switches, such as a color "reversal," or a conversion of all light colors to black, and all dark colors to white. (The screen captures given in this book were produced using the "reverse" color map.) Any such color changes must be redone when the program is restarted.

Other system parameters may likewise be set from the **File | Configuration** menu item. These include the path for temporary files that the program may create (or want to read), the mouse "double click" speed—important for those with slow reflexes—an added time delay to slow down programs on computers that are too fast, and a "check memory" option—primarily of interest to those making program modifications.

Those users wishing more information on the CUPS utilities should consult the CUPS Utilities Manual, written by Jaroslaw Tuszynski and William MacDonald, published by John Wiley and Sons. However, it is not necessary for casual users of CUPS programs to become familiar with the utilities. Such familiarity would only be important to someone wishing to write their own simulations using the utilities. The utilities are freely available for this purpose, for unrestricted noncommercial production and distribution of programs. However, users of the utilities who wish to write programs for commercial distribution should contact John Wiley and Sons.

1.5 Communicating With the Authors

Users of these programs should not expect that run-time errors will never occur! In most cases, such run-time errors may require only that the user restart the program; but in other cases, it may be necessary to reboot the computer, or even turn it off and on. The causes of such run-time errors are highly varied. In some cases, the program may be telling you something important about the physics or the numerical method. For example, you may be trying to use a numerical method beyond its range of applicability. Other types of run-time errors may have to do with memory or other limitations of your computer. Finally, although the programs in this book have been extensively tested, we cannot rule out the possibility that they may contain errors. (Please let us know if you find any! It would be most helpful if such problems were communicated by electronic mail, and with complete specificity as to the circumstances under which they arise.)

It would be best if you communicated such problems directly to the author of each program, and simultaneously to the editors of this book (the CUPS Directors), via electronic mail—see addresses listed below. Please feel free to communicate any suggestions about the programs and text which may lead to improvements in future editions. Since the programs have been provided in source code form, it will be possible for you to make corrections of any errors that you or we find in the future—provided that you send in the registration card at the back of the book, so that you can be notified. The fact that you have the source code will also allow you to make modifications and extensions of the programs. We can assume no responsibility for errors that arise in programs that you have modified. In fact, we strongly urge you to change the program name, and to add a documentary note at the beginning of the code of any modified programs that alerts other potential users of any such changes.

1.6 CUPS Courses and Developers

- **CUPS Directors**
  Maria Dworzecka, George Mason University (cups@gmuvax.gmu.edu)
  Robert Ehrlich, George Mason University (cups@gmuvax.gmu.edu)
  William MacDonald, University of Maryland (w_macdonald@umail.umd.edu)

- **Astrophysics**
  J. M. Anthony Danby, North Carolina State University (n38hs901@ncuvm.ncsu.edu)
  Richard Kouzes, Battelle Pacific Northwest Laboratory (rt_kouzes@pnl.gov)
  Charles Whitney, Harvard University (whitney@cfa.harvard.edu)

- **Classical Mechanics**
  Bruce Hawkins, Smith College (bhawkins@smith.bitnet)
  Randall Jones, Loyola College (rsj@loyvax.bitnet)

- **Electricity and Magnetism**
  Robert Ehrlich, George Mason University (rehrlich@gmuvax.gmu.edu)
  Lyle Roelofs, Haverford College (lroelofs@haverford.edu)
  Ronald Stoner, Bowling Green University (stoner@andy.bgsu.edu)
  Jaroslaw Tuszynski, George Mason University (cups@gmuvax.gmu.edu)

- **Modern Physics**
  Douglas Brandt, Eastern Illinois University (cfdeb@ux1.cts.eiu.edu)
  John Hiller, University of Minnesota, Duluth (jhiller@d.umn.edu)
  Michael Moloney, Rose Hulman Institute (moloney@nextwork.rose-hulman.edu)

- **Nuclear and Particle Physics**
  Roberta Bigelow, Willamette University (rbigelow@willamette.edu)
  John Philpott, Florida State University (philpott@fsunuc.physics.fsu.edu)
  Joseph Rothberg, University of Washington (rothberg@phast.phys.washington.edu)

- **Quantum Mechanics**
  John Hiller, University of Minnesota Duluth (jhiller@d.umn.edu)
  Ian Johnston, University of Sydney (idj@suphys.physics.su.oz.au)
  Daniel Styer, Oberlin College (dstyer@physics.oberlin.edu)

- **Solid State Physics**
  Graham Keeler, University of Salford (g.j.keeler@sysb.salford.ac.uk)
  Roger Rollins, Ohio University (rollins@chaos.phy.ohiou.edu)
  Steven Spicklemire, University of Indianapolis (steves@truevision.com)

- **Thermal and Statistical Physics**
  Harvey Gould, Clark University (hgould@vax.clarku.edu)
  Lynna Spornick, Johns Hopkins University
  Jan Tobochnik, Kalamazoo College (jant@kzoo.edu)

- **Waves and Optics**
  G. Andrew Antonelli, Wolfgang Christian, and Susan Fischer, Davidson College (wc@phyhost.davidson.edu)
  Robin Giles, Brandon University (giles@brandonu.ca)
  Brian James, Salford University (b.w.james@sysb.salford.ac.uk)

1.7 Descriptions of all CUPS Programs

Each of the computer simulations in this book (as well as those in the eight other books comprised by the CUPS Project) are described below. The individual headings under which programs appear correspond to the nine CUPS courses. In several cases, programs are listed under more than one course. The number of programs listed under the Astrophysics, Modern Physics, and Thermal Physics courses is appreciably greater than the others, because several authors have opted to subdivide their programs into many smaller programs. Detailed inquiries regarding CUPS programs should be sent to the program authors.

ASTROPHYSICS PROGRAMS

STELLAR (Stellar Models), written by Richard Kouzes, is a simulation of the structure of a static star in hydrodynamic equilibrium. This provides a model of a zero age main sequence star, and helps the user understand the physical processes that exist in stars, including how density, temperature, and luminosity depend on mass. Stars are self-gravitating masses of hot gas supported by thermodynamic processes fueled by nuclear fusion at their core. The model integrates the four differential equations governing the physics of the star to reach an equilibrium condition which depends only on the star's mass and composition.

EVOLVE (Stellar Evolution), written by Richard Kouzes, builds on the physics of a static star, and considers (1) how a gas cloud collapses to become a main sequence star, and (2) how a star evolves from the main sequence to its final demise. The model is based on the same physics as the STELLAR program. Starting from a diffuse cloud of gas, a protostar forms as the cloud collapses and reaches a sufficient density for fusion to begin. Once a star reaches equilibrium, it remains for most of its life on the main sequence, evolving off after it has consumed its fuel. The final stages of the star's life are marked by rapid and dramatic evolution.

BINARIES is the driver program for all Binaries Programs (**VISUAL1, VISUAL2, ECLIPSE, SPECTRO, TIDAL, ROCHE, and ACCRDISK**).

VISUAL1 (Visual Binaries—Proper Motion), written by Anthony Danby, enables you to visualize the proper motion in the sky of the members of a visual binary system. You can enter the elements of the system and the mass ratio, as well as the speed at which the center of mass moves across the screen. The program also includes an animated three-dimensional demonstration of the elements.

VISUAL2 (Visual Binaries—True Orbit), written by Anthony Danby, enables you to select an apparent orbit for the secondary star with arbitrary eccentricity, with the primary at any interior point. The elements of the orbit are displayed. You can see the orbit animated in three dimensions, or can make up a set of "observations" based on the apparent orbit.

ECLIPSE (Eclipsing Binaries), written by Anthony Danby, shows simultaneously either the light curve and the orbital motion or the light curve and an animation of the eclipses. You can select the elements of the orbit and radii and magnitudes of the stars. A form of limb-darkening is also included as an option.

SPECTRO (Spectroscopic Binaries), written by Anthony Danby, allows you to select the orbital elements of a spectroscopic binary, and then shows simultaneously the velocity curve, the orbital motion, and a moving spectral line.

TIDAL (Tidal Distortion of a Binary), written by Anthony Danby, models the motion of a spherical secondary star around a primary that is tidally distorted by the secondary. You can select orbital elements, masses of the stars, a parameter describing the tidal lag, and the initial rate of rotation of the primary. The equations are integrated over a time interval that you specify. Then you can see the changes of the orbital elements, and the rotation of the primary, with time. You can follow the motion in detail around each revolution, or in a form where the equations have been averaged around each revolution.

ROCHE (The Photo-Gravitational Restricted Problem of Three Bodies), written by Anthony Danby, follows the two-dimensional motion of a particle that is subject to the gravitational attraction of two bodies in mutual circular orbits, and also, optionally, radiation pressure from these bodies. It is intended, in part, as background for the interpretation of the formation of accretion disks. Curves of zero velocity (that limit regions of possible motion) can be seen. The orbits can also be followed using Poincaré maps.

ACCRDISK (Formation of an Accretion Disk), written by Anthony Danby, follows some of the dynamical steps in this process. The dynamics is valid up to the initial formation of a hot spot, and qualititative afterward.

NBMENU is the driver program for all programs on the motion of N interacting bodies: **TWOGALAX, ASTROIDS, N-BODIES, PLANETS, PLAYBACK, and ELEMENTS**.

TWOGALAX (The Model of Wright and Toomres), written by Anthony Danby, is concerned with the interaction of two galaxies. Each consists of a central gravitationally attracting point, surrounded by rings of stars (which are attracted, but do not attract). Elements of the orbits of one galaxy relative to the other are selected, as is the initial distribution and population of the rings. The motion can be viewed as projected into the plane of the orbit of the galaxies, or simultaneously in that plane and perpendicular to it. The positions can be stored in a file for later viewing.

ASTROIDS (N-Body Application to the Asteroids), written by Anthony Danby, uses the same basic model, but a planet and a star take the place of the galaxies and the asteroids replace the stars. Emphasis is on asteroids all having the same period, with interest on periods having commensurability with the period of the planet. The orbital motion of the system can be followed. The positions can be stored in a file for later viewing. An asteroid can be selected, and the variation of its orbital elements can then be followed.

NBODIES (The Motion of N Attracting Bodies), written by Anthony Danby, allows you to choose the number of bodies (up to 20) and the total energy of the system. Initial conditions are chosen at random, consistent with this energy, and the resulting motion can be observed. During the motion various quantities, such as the kinetic energy, are displayed. The positions can be stored in a file for later viewing.

PLANETS (Make Your Own Solar System), written by Anthony Danby, is similar to the preceding program, but with the bodies interpreted as a star with planets. Initial conditions are specified through the choice of the initial elements of the planets. The positions can be stored in a file for later viewing.

PLAYBACK, written by Anthony Danby, enables a file stored by one of the preceding programs to be viewed.

ELEMENTS (Orbital Elements of a Planet), written by Anthony Danby, shows a three-dimensional animation that can be viewed from any angle.

GALAXIES is the driver program for Galactic Kinematics Programs: **ROTATION, OORTCONS, and ARMS21CM**.

ROTATION (The Rotation Curve of a Galaxy), written by Anthony Danby, first prompts you to "design" a galaxy, consisting of a central mass and up to five spheroids (that can be visible or invisible). It then displays the galaxy and can show the animated rotation or the rotation curve.

OORTCONS (Galactic Kinematics and Oort's Constants), written by Anthony Danby, allows you to design your galaxy, choose the location of the "sun" and a local region around it, and the to observe the kinematics in this region. It also shows graphs of radial velocity and proper motion in comparison with the linear approximation, and computes the Oort constants.

ARMS21CM (The Spiral Structure of a Galaxy), written by Anthony Danby, allows you to design your galaxy, construct a set of spiral arms, and select the position of the "sun." Then, for different galactic longitudes, you can see observed profiles of 21 cm lines.

ATMOS (Stellar Atmospheres), written by Charles Whitney, permits the user to select a constellation, see it mapped on the computer screen, point to a star, and see it plotted on a brightness-color diagram. The user's task is to build a model atmosphere that imitates the photometric properties of observed stars. This is done by specifying numerical values for three basic stellar parameters: radius, mass, and luminosity. The program then builds the model and displays it on the brightness-color diagram, and it also plots the spectrum and the detailed thermodynamic structure of the atmosphere. With this program the user may investigate the relation between stellar parameters and the thermal properties of the gas in the atmosphere. Two atmospheres may be superposed on the graphs, for easier comparison.

PULSE (Stellar Pulsations), written by Charles Whitney, illustrates stellar pulsation by simulating the thermo-mechanical behavior of a "star" modelled by a self-gravitating gas divided by spherical elastic shells. The elastic shells resemble a set of coupled oscillators. The program solves for the modes of small-amplitude motion, and it uses Fourier synthesis to construct motions for arbitrary starting conditions. The screen displays the thermodynamic structure and surface properties, such as temperature, pressure, and velocity. Animation displays the nature of the pulsation. By showing the motions, temperatures, and energy flux, the program demonstrates the heat engine acting inside the pulsating star. The motions of the shells and the spatial Fourier decomposition into eigenmodes are displayed simultaneously, and this will help you visualize the meaning of the Fourier components.

CLASSICAL MECHANICS PROGRAMS

GENMOT (The Motion Generator), written by Randall Jones, allows you to solve numerically any differential equation of motion for a system with up to three degrees of freedom and display the time evolution of the system in a wide variety of formats. Any of the dynamical variables or any function of those variables may be displayed graphically and/or numerically and a wide range of animations may be constructed. Since the Motion Generator can be used to solve any second-order differential equation, it can also be used to study systems analyzed by Lagrangian methods. Real world coordinates may be constructed as functions of generalized coordinates so that simulations of the actual system can be constructed.

ROTATE (Rotation of 3-D Objects), written by Randall Jones, is designed to aid in the visualization of the dynamical variables of rotational motion. It will allow you to observe the 3-D motion of rotating objects in a controlled fashion, running the simulation faster, slower, or in reverse while displaying the corresponding evolution of the angular velocity, the angular momentum and the torque. It will display the motion from the fixed frame and from the body frame to help in understanding the translation between these two descriptions of the motion. By using the stereographic feature of the program you can create a genuine 3-D representation of the motion of the quantities.

COUPOSC (Coupled Oscillators), written by Randall Jones, is designed to investigate a wide range of harmonic systems. Given a set of objects and springs connected in one or two dimensions, the simulation can solve the problem by generating the normal mode frequencies and their corresponding motions. It can take any set of initial conditions and resolve them into their component normal mode motions or take any set of initial mode occupations and display the corresponding motions of the objects. It can also determine the motion of the system when it is acted on by external forces. In this case the total forces are no longer harmonic, so the solution is generated numerically. The harmonic analysis, however, still provides an important tool for investigating and understanding the subsequent motion.

ANHARM (Anharmonic Oscillators), written by Bruce Hawkins, simulates oscillations of various types: pendulum, simple harmonic oscillator, asymmetric, cubic, Vanderpol, and a mass in the center of a spring with fixed ends. Nonlinear behavior is emphasized. The user may choose to view one to four graphs of the motion simultaneously, along with the potential diagram and a picture of the moving object. Graphs that may be viewed are x vs. t, v vs. t, v vs. x, the Poincaré diagram, and the return map. Tools are provided to explore parameter space for regions of interest. Fourier analysis is available, resonance diagrams can be plotted, and the period can be plotted as a function of energy. Includes a tutorial demonstrating the usefulness of phase plots and Poincaré plots.

ORBITER (Gravitational Orbits), written by Bruce Hawkins, simulates the motion of up to five objects with mutually gravitational attraction, and any reasonable number of additional objects moving in the gravitation field of the first five. The motion may be viewed in up to six windows simultaneously: windows centered on a particular body, on the center of mass, stationary in the universe frame, or rotating with the line joining the two most massive bodies. A menu of available system includes the solar system, the sun/earth/moon system; the sun, Jupiter, and its moons; the sun, earth, and Saturn, demonstrating retrograde motion; the sun, Jupiter, and a comet; and a pair of binary stars with a comet. Bodies my be added to any system, or a new system created using either numerical coordinates or the mouse. Bodies may be replicated to demonstrate the sensitivity of orbits to initial conditions.

COLISION (Collisions), written by Bruce Hawkins, simulates two-body collisions under any of a number of force laws: Coulomb with and without shielding and truncation, hard sphere, soft sphere (harmonic), Yukawa, and Woods-Saxon. Collision may be viewed in the laboratory and center of mass systems simultaneously, with or without momentum diagrams. Includes a tutorial on the usefulness of the center of mass system, one on the kinematics of relativistic collisions, and one on cross section. Plots cross section against scattering parameter, and compares collisions at different parameters.

ELECTRICITY AND MAGNETISM PROGRAMS

FIELDS (Analysis of Vector and Scalar Fields), written by Jarek Tuszynski, displays scalar and vector fields for any algebraic or trigonometric expression entered by the user. It also computes numerically the divergence, curl, and Laplacian for the vector fields, and the gradient and Laplacian for the scalar fields. Simultaneous displays of selected quantities are provided in user-selected planes, using vector, contour, or 3-D plots. The program also allows the user to define paths along which line integrals are computed.

GAUSS (Gauss' Law), written by Jarek Tuszynski, treats continuous charge distributions having spherical or cylindrical symmetry, and those that vary as a function of the x-coordinate only. The program allows the user to enter an arbitrary function to define either the electric field magnitude, the potential, or the charge density. It then computes the other two functions by numerical differentiation or integration, and displays all three functions. Finally, the program allows the user to enter a "comparison function," which is plotted on the same graph, so as to check whether his analytic solutions are correct.

POISSON (Poisson's Equation Solved on a Grid), written by Jarek Tuszynski, solves Poisson's equation iteratively on a 2-D grid using the method of simultaneous over-relaxation. The user can draw arbitrary systems consisting of line charges, and charged conducting cylinders, plates, and wires, all infinite in extent perpendicular to the grid. After iteratively solving Poisson's equation, the program displays the results for the potential, electric field, or the charge density (found from the Laplacian of the potential), in the form of contour, vector, or 3-D plots. In addition, many other program features are available, including the ability to specify surfaces, along which the potential varies according to some algebraic function specified by the user.

IMAG&MUL (Image Charges and Multipole Expansion), written by Lyle Roelofs and Nathaniel Johnson, allows users to explore two approaches to the solution of Laplace's equation—the image charge method and expansion in multipole moments. In the image charge mode (IC) the user is presented with a variety of configurations involving conducting planes and point charges and is asked to "solve" each by placing image charges in the appropriate locations. The program displays the electric field due to all point charges, real and image, and a solution can be regarded as successful with the field due to all charges is everywhere orthogonal to all conducting surfaces. Solutions can then be examined with a variety of included software "tools." The multipole expansion (ME) mode of the program also permits a "hands-on" exploration of standard electrostatic problems, in this case the "exterior" problem, i.e., the determination of the field outside a specified equipotential surface. The program presents the user with a variety of azimuthally symmetric equipotential surfaces. The user "solves" for the full potential by adding chosen amounts of the (first six) multipole moments. The screen shows the contours of the summed potential and the problem is "solved" when the innermost contour matches the given equipotential surface as closely as possible.

ATOMPOL (Atomic Polarization), written by Lyle Roelofs and Nathaniel Johnson, is an exploration of the phenomenon of atomic polarization. Up to 36 atoms of controllable polarizibility are immersed in an external electric field. The program solves for and displays the field throughout the region in which the atoms are located. A closeup window shows the polarization of selected atoms and software "tools" allow for further analysis of the resulting electric fields. Use of this program improves the student's understanding of polarization, the interaction of polarized entities and the atomic origin of macroscopic polarization, the latter via study of closely spaced clusters of polarizable atoms.

DIELECT (Dielectric Materials), written by Lyle Roelofs and Nathaniel Johnson, is a simulation of the behavior of linear dielectric materials using a cell-based approach. The user controls either the polarization or the susceptibility of each cell in a (25 $\times$ 25) grid (with assume uniformity in the third direction). Full self-consistent solutions are obtained via an iterative relaxation method and the fields P, E, or D are displayed. The student can investigate the self-interactions of polarized materials and many geometrical effects. Use of this program aids the student in developing understanding of the subtle relations among and meaning of P, E, and D.

ACCELQ (Fields From an Accelerated Charge), written by Ronald Stoner, simulates the electromagnetic fields in the plane of motion generated by a point charge that is moving and accelerating in two dimensions. The user chooses from among seven predefined trajectories, and sets the values of maximum speed and viewing time. The electric field pattern is recomputed after each change of trajectory or parameter; thereafter, the user can investigate the electric field, magnetic field, retarded potentials, and Poynting-vector field by using the mouse as a field probe, by using gridded overlays, or by generating plots of the various fields along cuts through the viewing plane.

QANIMATE (Fields From an Accelerated Charge—Animated Version), written by Ronald Stoner, is an interactive animation of the changing electric field pattern generated by a point electric charge moving in two dimensions. Charge motion can be manipulated by the user from the keyboard. The display can include electric field lines, radiation wave fronts, and their points of intersection. The motion of the charge is controlled by the using **arrow** keys to accelerate and steer much like the accelerator and steering wheel of a car, except that acceleration must be changed in increments, and the **Space** bar can used to engage or disengage the steering. With steering engaged, the charge will move in a circle. Unless the acceleration is made zero, the speed will increase (or decrease) to the maximum (minimum) possible value. At constant speed and turning rate, the charge can be controlled by the **Space** bar alone.

EMWAVE (Electromagnetic Waves), written by Ronald Stoner, uses animation to illustrate the behavior of electric and magnetic fields in a polarized plane electromagnetic wave. The user can choose to observe the wave in free space, or to see the effect on the wave of incidence on a material interface, or to see the effects of optical elements that change its polarization. The user can change the polarization state of the incident wave by specifying its Stokes parameters. Standing electromagnetic waves can be simulated by combining the incident travelling wave with a reflected wave of the same amplitude. The user can do that by choosing appropriate values of the physical properties of the medium on which the incident wave impinges in one of the animations.

MAGSTAT (Magnetostatics), written by Ronald Stoner, computes and displays magnetic fields in and near magnetized materials. The materials are uniform and have 3-D shapes that are solids of revolution about a vertical axis. The shape of the material can be modified or chosen from a data input screen. The user has the option of generating the fields produced by a permanently and uniformly magnetized object, or of generating the fields of a magnetizable object placed in an otherwise uniform external field. Besides choosing the shape and aspect ratio of the object, the user can vary the magnetic permeability of the magnetizable material, and choose among three fields to display: magnetic induction (B), magnetic field strength (H), and magnetization (M). Each of these fields can be displayed or explored in several different ways. The algorithm for computing the fields uses a superposition of Chebyschev polynomial approximants to the H field due to "rings" of "magnetic charge."

MODERN PHYSICS PROGRAMS

NUCLEAR (Nuclear Energetics and Nuclear Counting), written by Michael Moloney, deals with basic nuclear properties related to mass, charge, and energy, for approximately 1900 nuclides. Graphs are available involving binding energy, mass, and Q values of a variety of nuclear reactions, including alpha and beta decays. Part 2 deals with simulating the statistics of counting with a Geiger-Muller tube. This part also simulates neutron activation, and the counting behavior as neutron flux is turned on and off. Finally, a decay chain from A to B to C is simulated, where half-lives may be changed, and populations are graphed as a function of time.

GERMER (Davisson-Germer and G. P. Thomson Experiments), written by Michael Moloney, simulates both the Davisson-Germer and G. P. Thomson experiments with electrons scattering from crystalline materials. Stress is laid on the behavior of electrons as waves; similarities are noted with scattering of x-rays. The exercises encourage students to understand why peaks and valleys in scattered electrons occur where they do.

QUANTUM (one-dimensional Quantum Mechanics), written by Douglas Brandt, is a program that has four sections. The first section allows users to investigate the uncertainty principle for specified wavefunctions in position or momentum space. The second section allows users to investigate the time evolution of wavepackets under various dispersion relations. The third section allows users to investigate solutions to Schrödinger's equation for asymptotically free solutions. The user can input a barrier and the program calculates reflection and transmission coefficients for a range of energies and show wavepacket time evolution for the barrier potential. The fourth section is similar to the third, except that it allows the user to investigate bound solutions to Schrödinger's equation. The program calculates the bound state Hamiltonian eigenvalues and spatial eigenfunctions.

RUTHERFD (Rutherford Scattering), written by Douglas Brandt, is a program for investigating classical scattering of particles. A scattering potential can be chosen from a list of predefined potentials or an arbitrary potential can be input by the user. The computer generates scattering events by randomly picking impact parameters from a distribution defined by beam parameters specified by the user. It displays the results of the scattering on a polar histogram and on a detailed histogram to help users gain insight into differential scattering cross section. A scintillation mode can be chosen for users that want more appreciation of the actual experiments of Geiger and Marsden. A "guess the scatterer" mode is available for trying to gain appreciation of how scattering experiments are used to infer properties of the scatterers.

SPECREL (Special Relativity), written by Douglas Brandt, is a program to investigate special relativity. The first section is to investigate change of coordinate systems through Minkowski diagrams. The user can define coordinates of objects in one reference frame and the computer calculates the coordinates in a user-selectable coordinate system and displays the objects in both reference frames. The second section allows users to view clocks that are in relative motion. A clock can be given an arbitrary trajectory through space-time and the readings of various clocks can be viewed as the clock follows that trajectory. A third section allows users to observe collisions in different reference frames that are related by Lorentz transformations or by Gallilean transformations.

LASER (Lasers), written by Michael Moloney, simulates a three-level laser, with the user in control of energy level parameters, temperature, pump power, and end mirror transmission. Atomic populations may be graphically tracked from thermal equilibrium through the lasing threshold. A mirror cavity simulation is available which uses ray tracing. This permits study of cavity stability as a function of mirror shape and position, as well as beam shape characteristics within the cavity.

HATOM (Hydrogenic Atoms), written by John Hiller, computes eigenfunctions and eigenenergies for hydrogen, hydrogenic atoms, and single-electron diatomic ions. Hydrogenic atoms may be exposed to uniform electric and magnetic fields. Spin interactions are not included. The magnetic interaction used is the quadratic Zeeman term; in the absence of spin-orbit coupling, the linear term adds only a trivial energy shift. The unperturbed hydrogenic eigenfunctions are computed directly from the known solutions. When external fields are included, approximate results are obtained from basis-function expansions or from Lanczos diagonalization. In the diatomic case, an effective nuclear potential is recorded for use in calculation of the nuclear binding energy.

NUCLEAR AND PARTICLE PHYSICS PROGRAMS

NUCLEAR (Nuclear Energetics and Counting), written by Michael Moloney, is included here, but is described under the Modern Physics Heating.

SHELLMOD (Nuclear Models), written by Roberta Bigelow, calculates energy levels for spherical and deformed nuclei using the single particle shell model. You can explore how the nuclear potential shape, the spin-orbit interaction, and deformation affect both the order and spacing of nuclear energy levels. In addition, you will learn how to predict spin and parity for single particle states.

NUCRAD (Interaction of Radiation With Matter), written by Roberta Bigelow, is a simulation of alpha particles, muons, electrons, or photons interacting with matter. You will develop an understanding of how ranges, energy losses, and random particle paths depend on materials, radiation, and incident energy. As a specific application, you can explore photon and electron interactions in a sodium iodide crystal which determines the energy response of a radiation detector.

ELSCATT (Electron-Nucleus Scattering), by John Philpott, is an interactive software tool that demonstrates various aspects of electron scattering from nuclei. Specific features include the relativistic kinematics of electron scattering, densities and form factors for elastic and inelastic scattering, and the nuclear Coulomb response. The simulation illustrates how detailed nuclear structure information can be obtained from electron scattering measurements.

TWOBODY (Two-Nucleon Interactions), by John Philpott, is an interactive software tool that illuminates many features of the two-nucleon problem. Bound state wavefunctions and properties can be calculated for a variety of interactions that may include non-central parts. Phase shifts and cross sections for pp, pn, and nn scattering can be calculated and compared with those obtained experimentally. Spin-polarization features of the cross sections can be extensively investigated. The simulation demonstrate the richness of the two-nucleon data and its relation to the underlying nucleon-nucleon interaction.

RELKIN (Relativistic Kinematics), by Joseph Rothberg, is an interactive program to permit you to explore the relativistic kinematics of scattering reactions and two-body particle decays. You may choose from among a large number of initial and final states. The initial momentum of the beam particle and the center of mass angle of a secondary can also be specified. The program displays the final state vector momenta in both the lab system and center of the mass system along with numerical values of the most important kinematic quantities. The program may be run in a Monte Carlo mode, displaying a scatter plot and histogram of selected variables. The particle data base may be modified by the user and additional reactions and decay modes may be added.

DETSIM (Particle Detector Simulation), by Joseph Rothberg, is an interactive tool to allow you to explore methods of determining parameters of a decaying particle or scattering reaction. The program simulates the response of high-energy particle detectors to the final-state particles from scattering or decays. The detector size and location may be specified by the user as well as its energy and spatial resolution. If the program is run in a Monte Carlo mode, detector hit information for each event is written to a file. This file can be read by a small reconstruction and plotting program. You can easily modify one of the example reconstruction programs that are provided to determine the mass, momentum, and other properties of the initial particle or state.

QUANTUM MECHANICS PROGRAMS

BOUND1D (Bound States in One Dimension), written by Ian Johnston, is a tool which allows you to explore energy eigenfunctions for an electron in various potential wells, which can be square, parabolic, ramped, asymmetric, double or Coulombic. The first part of the program deals with finding the eigenvalues and eigenfunctions of different wells. You may find them yourself, using a "hunt and shoot" method, or else the program will compute the eigenvalues automatically, by counting the number of nodes to determine where the eigenvalues occur. The second part of the program looks at properties of eigenfunctions normalization, orthogonality and the evaluation of many kinds of overlap integrals. The third part examines time development of general states made up of a superposition of bound state eigenfunctions. Facility is provided for you to incorporate your own procedures to specify different potential wells or different overlap integrals.

SCATTR1D (Scattering in One Dimension), written by John Hiller, solves the time-independent Schrödinger equation for stationary scattering states in one-dimensional potentials. The wavefunction is displayed in a variety of ways, and the transmission and reflection probabilities are computed. The probabilities may be displayed as functions of energy. The computations are done by numerically integrating the Schrödinger equation from the region of the transmitted wave, where the wavefunction is known up to some overall normalization and phase, to the region of the incident wave. There the reflected and incident waves are separated. The potential is assumed to be zero in the incident region and constant in the transmitted region.

QMTIME (Quantum Mechanical Time Development), written by Daniel Styer, simulates quantal time development in one dimension. A variety of initial wave packets (Gaussian, Lorentzian, etc.) can evolve in time under the influence of a variety of potential energy functions (step, ramp, square well, harmonic oscillator, etc.) with or without an external driving force. A novel visualization technique simultaneously displays the magnitude and phase of complex-valued wave functions. Either position-space or momentum-space wave functions, or both, can be shown. The program is particularly effective in demonstrating the classical limit of quantum mechanics.

LATCE1D (Wavefunctions on a one-dimensional Lattice), written by Ian Johnston, is a tool which allows you to explore energy eigenfunctions for an electron in a lattice made up of a number of simple potential wells (up to twelve), which can be square, parabolic, or Coulombic. You may find the eigenvalues yourself, using a "hunt and shoot" method, or allow the program to compute them automatically. You can firstly explore regular lattices, where all wells are the same and spaced at regular intervals. These will demonstrate many of the properties of regular crystals, particularly the existence of energy bands. Secondly you can change the width, depth or spacing of any of the wells, which will mimic the effect of impurities or other irregularities in a crystal. Lastly you can apply an external electric across the lattice. Facility is provided for you to incorporate your own procedures to calculate wells, lattice arrangements or external fields of their own choosing.

BOUND3D (Bound States in Three Dimensions), written by Ian Johnston, is a tool which allows you to explore energy eigenfunctions for an particle in a spherically symmetric potential well, which can be square, parabolic, Coulombic, or several other shapes of importance in molecular or nuclear applications. The first part of the program deals with finding the eigenvalues and eigenfunctions of different wells, assuming that the angular part of the wavefunctions are spherical harmonics. You may find them yourself for a given angular momentum quantum number using a "hunt and shoot" method, or else the program will compute the eigenvalues automatically, by counting the number of nodes to determine where the eigenvalues occur. The second part of the program looks at properties of eigenfunctions normalization, orthogonality and the evaluation of many kinds of overlap integrals. Facility is provided for you to incorporate your own procedures to specify different potential wells or different overlap integrals.

IDENT (Identical Particles in Quantum Mechanics), written by Daniel Styer, shows the probability density associated with the symmetrized, antisymmetrized, or nonsymmetrized wave functions of two noninteracting particles moving in a one-dimensional infinite square well. It is particularly valuable for demonstrating the effective interaction of noninteracting identical particles due to interchange symmetry requirements.

SCATTR3D (Scattering in Three Dimensions), written by John Hiller, performs a partial-wave analysis of scattering from a spherically symmetric potential. Radial and 3-D wave functions are displayed, as are phase shifts, and differential and total cross sections. The analysis employs an expansion in the natural angular momentum basis for the scattering wavefunction. The radial wavefunctions are computed numerically; outside the region where the potential is important they reduce to a linear combination of Bessel functions which asymptotically differs from the free radial wavefunction by only a phase. Knowledge of these phase shifts for the dominant values of angular momentum is used to approximate the cross sections.

CYLSYM (Cyllindrically Symmetric Potentials), written by John Hiller, solves the time-independent Schrödinger equation Hu=Eu in the case of a cylindrically symmetric potential for the lowest state of a chosen parity and magnetic quantum number. The method of solution is based on evolution in imaginary time, which converges to the state of the lowest energy that has the symmetry of the initial guess. The Alternating Direction Implicit method is used to solve a diffusion equation given by $HU = -\hbar \partial U/\partial t$, where H is the Hamiltonian that appears in the Schrödinger equation. At large times, U is nearly proportional to the lowest eigenfunction of H, and the expectation value $\langle H \rangle = \langle U | H | U \rangle / \langle U | U \rangle$ is an estimate for the associated eigenenergy.

SOLID STATE PHYSICS

LATCE1D (Wavefunctions for a one-dimensional Lattice), written by Ian Johnston, and included here, is described under the Quantum Mechanics heading.

SOLIDLAB (Build Your Own Solid State Devices), written by Steven Spicklemire, is a simulation of a semiconductor device. The device can be "drawn" by the user, and the characteristics of the device adjusted by the user during the simulation. The user can see how charge density, current density, and electric potential vary throughout the device during its operation.

LCAOWORK (Wavefunctions in the LCAO Approximation), written by Steven Spicklemire, is a simulation of the interaction of 2-D atoms within small atomic clusters. The atoms can be adjusted and moved around while their quantum mechanical wavefunctions are calculated in real time. The student can investigatge the dependence of various properties of these atomic clusters on the properties of individual atoms, and the geometric arrangement of the atoms within the cluster.

PHONON (Phonons and Density of States), written by Graham Keeler, calculates and displays phonon dispersion curves and the density of states for a number of different 3-D cubic crystal structures. The displays of the dispersion curves show realistic curves and allow the user to study the effect of changing the interatomic forces between nearest and further neighbor atoms and, for diatomic crystal structures, changing the ratio of the atomic masses. The density of states calculation shows how the complex shapes of real densities of states are built up from simpler distributions for each mode of polarization, and enables the user to match the features of the distribution to corresponding features on the dispersion curves. In order to help with visualization of the crystal lattices involved, the program also shows 3-D projections of the different crystal structures.

SPHEAT (Calculations of Specific Heat), written by Graham Keeler, calculates and displays the temperature variation of the lattice specific heat for a number of different theoretical models, including the Einstein model and the Debye model. It also makes the calculation for a computer simulation of a realistic density of states, in which the user can vary the important parameters of the crystal, including those affecting the density of state. The program can display the results for a small region near the origin, and as a T-cubed plot to enable the user to investigae the low temperature limit of the specific heat, or in the form of the equivalent Debye temperature to enhance a study of the deviations from the Debye model. The Schottky specific heat anomaly can also be investigated.

BANDS (Energy Bands), written by Roger Rollins, calculates and displays, for easy comparison, the energy dispersion curves and corresponding wavefunctions for an electron in a 1-D symmetric $V(x) = V(-x)$ periodic potential of arbitrary shape and of strength V_0. The method used is based on an exact, non-perturbative approach so that the energy dispersion curves and band gaps can be obtained for large V_0. Wavefunctions can be displayed, and compared with one another, by clicking the mouse on the desired states on the energy dispersion curve. Changes in band strtucture can be followed as changes are made in the shape of the potential. The variation of the band gaps with V_0 is calculated and compared with the two opposite limits of very weak V_0 (perturbation method) and very strong V_0 (isolated atom). Even the experienced condensed matter researcher may be surprised by some of the results! Open-ended class discussions can result from the interesting physics found in these conceptually simple model calculations.

PACKET (Electron Wavepacket in a 1-D Lattice), written by Roger Rollins, shows a live animation, calculated in real time, demonstrating how an electron wavepacket in a metal or semiconducting crystal moves under the influence of external forces. The time-dependent Schrödinger equation is solved in a tight binding approximation, including the external force terms, and the motion of the wavepacket is obtained directly. The main objective of the simulation is to show that an electron wavepacket formed from states with energies near the top of an energy band is accelerated in a direction *opposite* to the direction of the external force; it has a *negative* effective mass! The simulation deals with motion in a 1-D lattice but the concepts are applicable to the full 3-D motion of an electron in a real crystal. Numerical experiments on the motion of the packet explore interesting physics questions such as: how does constant applied force affect the periodic motion of a packet? when does the usual semiclassical model fail? what happens to the dynamics of the packet when placed in a superlattice with lattice constant twice that of the original lattice?

THERMAL AND STATISTICAL PHYSICS PROGRAMS

ENGDRV, written by Lynna Spornick, is a driver program for **ENGINE, DIESEL, OTTO, and WANKEL**. These programs provide an introduction to the thermodynamics of engines.

ENGINE (Design Your Own Engine), written by Lynna Spornick, lets the user design an engine by specifying the processes (adiabatic, isobaric, isochoric [constant volume], and isothermic) in the engine's cycle, the engine type (reversible or irreversible), and the gas type (helium, argon, nitrogen, or steam). The thermodynamic properties (heat exchanged, work done, and change in internal energy) for each process and the engine's efficiency are computed.

DIESEL, OTTO, and WANKEL, written by Lynna Spornick, provide animations of each of these types of engine. Plots of the temperature versus entropy and the pressure versus volume for the cycles are show with the engine's current thermodynamic conditions indicated.

PROBDRV, written by Lynna Spornick, is a driver program for **GALTON, POISEXP, TWOD, KAC, and STADIUM**. Subprograms GALTON, POISEXP, and TWOD provide an introduction to probability and subprograms KAC and STADIUM provide an introduction to statistics.

GALTON (A Galton Board), written by Lynna Spornick, models either a traditional Galton Board or a customized Galton Board with traps, reflecting, and/or absorbing walls. GALTON demonstrates the binominal and normal distributions, the laws of probability, and the central limit theorem.

POISEXP (Poisson Probability Distribution in Nuclear Decay), written by Lynna Spornick, uses the decay of radioactive atoms to describe the Poisson and the exponential distributions.

TWOD (2-D Random Walk), written by Lynna Spornick, models a random walk in two dimensions. A "drunk," taking equal-length steps, is required to walk either on a grid or on a plane. TWOD demonstrates the joint probability of two independent processes, the binominal distribution, and the Rayleigh distribution.

KAC (A Kac Ring), written by Lynna Spornick, uses a Kac ring to demonstrate that large mechanical systems, whose equations of motion are solvable and which obey time reversal and have a Poincaré cycle, can also be described by statistical models.

STADIUM (The Stadium Model), written by Lynna Spornick, uses a stadium model to demonstrate that there exists mechanical systems whose equations of motion are solvable but whose motion is not predictable because of the system's chaotic nature.

ISING (Ising Model in One and Two Dimensions), written by Harvey Gould, allows the user to explore the static and dynamic properties of the 1- and 2-D Ising model using four different Monte Carlo algorithms and three different ensembles. The choice of the Metropolis algorithm allows the user to study the Ising model at constant temperature and external magnetic field. The orientation of the spins is shown on the screen as well as the evolution of the total energy or magnetization. The mean energy, magnetization, heat capacity, and susceptibility are monitored as a function of the number of configurations that are sampled. Other computed quantities include the equilibrium-averaged energy and magnetization autocorrelation functions and the energy histogram. Important physical concepts that can be studied with the aid of the program include the Boltzmann probability, the qualitative behavior of systems near critical points, critical exponents, the renormalization group, and critical slowing down. Other algorithms that can be chosen by the user correspond to spin exchange dynamics (constant magnetization), constant energy (the demon algorithm), and single cluster Wolff dynamics. The latter is particularly useful for generating equilibrium configurations at the critical point.

MANYPART (Many Particle Molecular Dynamics), written by Harvey Gould, allows the user to simulate a dense gas, liquid, or solid in two dimensions using either molecular dynamics (constant energy, constant volume) or Monte Carlo (constant temperature, constant volume) methods. Both hard disks and the Lennard-Jones interaction can be chosen. The trajectories of the particles are shown as the system evolves. Physical quantities of interest that are monitored include the pressure, temperature, heat capacity, mean square displacement, distribution of the speeds and velocities, and the pair correlation function. Important physical concepts that can be studied with the aid of the program include the Maxwell-Boltzmann probability distribution, fluctuations, equation of state, correlations, and the importance of chaotic mixing.

FLUIDS (Thermodynamics of Fluids), written by Jan Tobochnik, allows the user to explore the fluid (gas and liquid) phase diagrams for the van der Waals model and water. The user chooses four phase diagrams from among the following choices: PT, Pv, vT, uT, sT, uv, and sv, where P is the pressure, T is the temperature, v is the specific volume, S is the specific entropy, and u is the specific internal energy. The program reads in the coexistence table for the van der Waals model and water, and uses it along with an empirical formula for the water free energy and the free energy derived from the van der Waals model. Given v and u, any thermodynamic quantity can be calculated. For the van der Waals model thermodynamic quantities also can be calculated from the other thermodynamic state variables. The user can draw a straight line path in one phase diagram and see how this path looks in the other phase diagrams. The user also can extract all important thermodynamic data at any point in a phase diagram.

QMGAS1 (Quantum Mechanical Gas—Part 1), written by Jan Tobochnik, does the numerical calculations necessary to solve for the thermodynamic properties of quantum ideal gases, including photons in blackbody radiation, ideal bosons, phonons in the Debye theory, non-interacting fermions, and the classical limits of these systems. The user chooses the type of statistics (Bose-Einstein, Fermi-Dirac, or Maxwell-Boltzmann), the dimension of space, the form of the dispersion relation (restricted to simple powers), whether or not the particles have a non-zero chemical potential, and whether or not there is a Debye cutoff. The program then allows the user to build up a table of thermodynamic data, including the energy, specific heat, and chemical potential as a function of temperature. This data and various distribution functions and the density of states can be plotted.

QMGAS2 (Quantum Mechanical Gas—Part2), written by Jan Tobochnik, implements a Monte Carlo simulation of a finite number of quantum particles fluctuating between various states in a finite k-space (k is the wavevector). The program orders the possible energy states into an energy level diagram and then allows particles to move from one state to another according to the usual Boltzman probability distribution. Bosons are restricted so that they may not pass through each other on the energy level diagram; fermions are further restricted so that no two fermions may be in the same state; classical particles have no restrictions. In this way indistinguishability is correctly implemented for bosons and fermions. The user chooses the type of particle, the number of particles, the size and dimension of k-space, and the temperature. During the simulation the user sees a representation of the state occupancy and plots of the average energy, the instantaneous energy, and the distribution of energy among the states, also shown are results for the average energy, specific heat, and the occupancy of the ground state.

WAVES AND OPTICS PROGRAMS

DIFFRACT (Interference and Diffraction), by Robin Giles, simulates some of the fundamental wave behaviors in Fresnel and Fraunhofer Diffraction, and other Interference and Coherence effects. In particular you will be able to study diffraction phenomena associated with a point or a set of points and a slit or set of slits using the Huyghens construction. You can also use a method developed by Cornu—the Cornu Spiral—to examine diffraction from one or two slits or one or two obstacles. You can study Fresnel and Fraunhofer diffraction with a single slit or set of slits, a rectangular aperture and a circular aperture. Finally you can study Partial Coherence and fringe visibility in interference and diffraction observations. In the latter example you will be able to study the Michelson Stellar Interferometer and measure the separation distance in a double star and measure the diameter of single stars.

SPECTRUM (Applications of Interfence and Diffraction), by Robin Giles, simulates the uses and modes of operation of four important optical instruments—the Diffraction Grating, the Prism Spectrometer, the Michelson Interferometer and the Fabry-Perot Interferometer. You will look at the nature of the spectra, simulated interference patterns, and the question of resolving power.

WAVE (One-Dimensional Waves), by Wolfgang Christian, Andrew Antonelli, and Susan Fischer, uses finite difference methods to study the time evolution of the following partial differential equations: classical wave, Schrödinger, diffusion, Klein-Gordon, sine-Gordon, phi four, and double sine-Gordon. The user may vary the initial function and boundary conditions. Unique features of the program include mouse-driven drawing tools that enable the user to create sources, segments, and detectors anywhere inside the medium. Double-clicking on a segment allows the user to edit properties such as index of refraction or potential in order to model barrier problems such as thin film interference filters or the Ramsauer-Townsend effect in optics and quantum mechanics, respectively. Various types of analysis can be performed, including detector value, space-time, Fourier analysis and energy density.

CHAIN (One-Dimensional Lattice of Coupled Oscillators), by Wolfgang Christian, Andrew Antonelli, and Susan Fischer, allows the user to examine the time evolution of a 1-D lattice of coupled oscillators. These oscillators are represented on screen as a chain of masses undergoing vertical displacement. The program allows the user to examine how the application of Newtonian mechanics to these masses leads to traveling and standing waves. The relationship between the lattice spacing and other properties such as dispersion, band gaps, and cut-off frequency can be examined. Each mass can be assigned linear, quadratic, and cubic nearest neighbor interactions as well as a time-dependent external force function. Global properties such as the total energy in the lattice or the Fourier transform of the lattice can be displayed as well as the time evolution of a single mass's dynamical variables.

FOURIER (Fourier Analysis and Synthesis), written by Brian James, allows investigation of Fourier analysis and 1-D and 2-D Fourier transforms. In Fourier analysis users can choose from several predefined functions or enter their own functions either algebraically, numerically or graphically. The build-up of a periodic function is illustrated as successive terms of the Fourier series are added in, and the effects of dispersion and attenuation on the evolution of the synthesized waveform can then be investigated. One- and two-dimensional discrete Fourier transforms can be produced for a range of standard and user-entered functions. The effects of filters on the inverse transforms are illustrated. The 2-D transforms are shown as surface and contour plots. Image processing can be illustrated by filtering the transforms of gray level images so that when the inverse transforms are displayed it can be seen that the images have been modified.

RAYTRACE (Ray Tracing and Lenses), by Brian James, lets the user explore the applications of ray tracing in geometrical optics. The fundamental principle of Fermat can be illustrated by plotting the path of a ray through two different materials between fixed points. The variation of the path of a ray through a region of changing refractive index can be used to investigate the formation of mirages. The variation of pulse delay in a fiber can be calculated as a function of its parameters and the characteristics of optical communication fibers are considered. The formation of primary and secondary rainbows due to dispersion of refractive index can be displayed. The matrix method of tracing rays through lenses can be used to investigate the images formed and show how aberrations in images arise and may be reduced.

QUICKRAY (Quick Ray Tracing), by John Philpott, can be used to demonstrate ray diagrams for a single thin lens or spherical mirror. The object and image are shown, along with the three principal rays that proceed from the object towards the observer. You can use the mouse to move the object, the position of the lens or mirror or to change the focal length of the lens or mirror. The principal rays are continuously redrawn while any of these adjustments are made. The simulation handles converging and diverging lenses and concave and convex mirrors. Thus students can quickly get an intuitive feel for real and virtual image formation under a variety of circumstances.

Acknowledgments

The CUPS Project was funded by the National Science Foundation (under grant number PHY-9014548), and it has received support from the IBM Corporation, the Apple Corporation, and George Mason University.

2

Thermodynamics of Fluids

Jan Tobochnik

2.1 Introduction

The program FLUIDS is designed to help you understand phase diagrams of the gas and liquid states of matter. In particular, you will be able to relate changes in one thermodynamic variable, such as temperature, to changes in another variable, such as pressure. In the following section we discuss the necessary background in thermodynamics and relate the concepts to the computational procedures used by the program FLUIDS.

2.2 Background

2.2.1 Phase Diagrams

Because matter can exist in many phases, it is useful to plot the boundaries of the phases as a function of two or three thermodynamic variables. Such plots are called *phase diagrams*. Two or more phases can coexist at the same value of the thermodynamic variables. For example, a closed container of water molecules will typically contain both liquid water and the gaseous or vapor form of water. The liquid phase will be at the bottom of the container and the vapor at the top. If gravity were not present, the two phases would not be separated in this way, and instead there would be droplets of liquid suspended in the vapor and bubbles of gas suspended in the liquid.

In general, a fluid exists as either a pure gas, a pure liquid, or both phases in coexistence. Above a certain temperature, called the critical temperature, and a pressure, called the critical pressure, there is no distinction between liquid and gas. The phase diagram shows the boundaries among these three possibilities.[1,2] Two-dimensional phase diagrams will show the phase boundaries as functions of a pair of thermodynamic variables. For example, a p-V diagram shows the phases as a function of pressure and volume. In a P-T diagram the region of coexistence is a curve with positive slope which ends at the critical point. On the low-temperature side of this curve is the liquid, and on the high-temperature side is the gas. Figure 2.1 shows four phase diagrams for water created by the program FLUIDS. One of the features of the program FLUIDS is that the user can draw a thermodynamic path in one phase diagram and the program will produce the path in the other phase diagrams. Figure 2.1 shows the thermodynamic paths calculated from a straight line segment drawn by the user in the V-T diagram.

One of the more interesting questions is, how do substances change from one phase to another, particularly near the critical point? Within the last two decades we have learned much about the behavior of thermodynamic quantities near the critical point. Very close to the critical point many thermodynamic quantities diverge with power laws whose form is independent of the type of fluid. There are critical exponents that measure the rate of divergence for each thermodynamic quantity, and these exponents are related to each other by what are called *scaling laws*. Another key idea, called *universality,* is that the exponents are the same for fluids and other types of phase transitions, such as the Curie point for ferromagnets.

The two phase coexistence region also is of much theoretical and practical interest. In particular, there is much interest in the nonequilibrium behavior of systems that are quenched or cooled rapidly into the two-phase region. Here the interest centers on the rate of nucleation of one phase within the other, and the structure of the two phases as a function of time when the quench occurs at the critical density. An understanding of phase diagrams and the thermodynamic properties of matter is essential background for working in these active research areas.

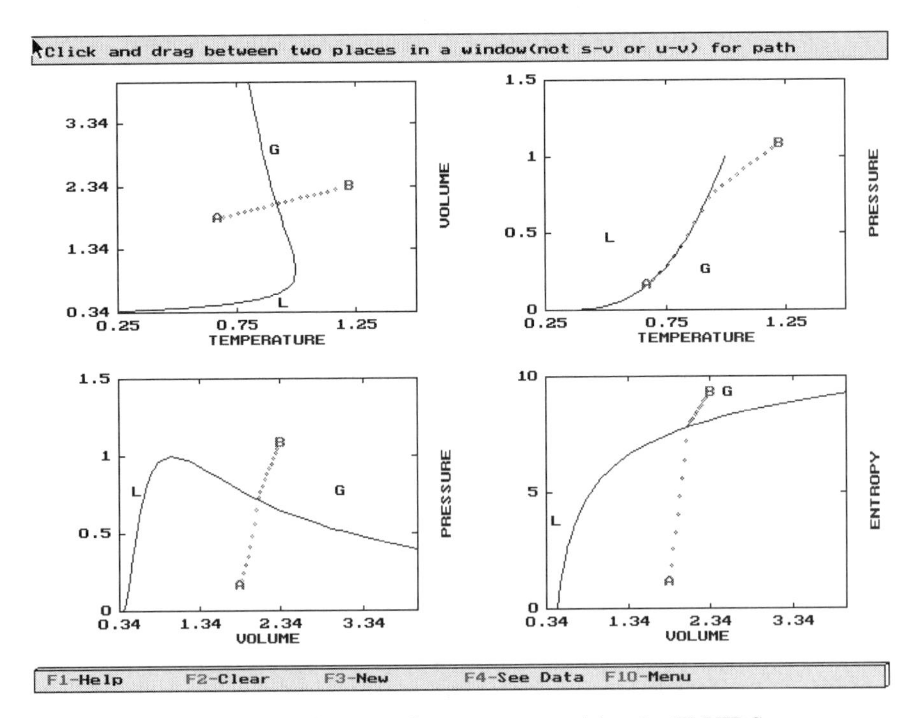

Figure 2.1: Four phase diagrams for water created by the FLUIDS program.

2.2.2 Helmholtz Free Energy

Vast amounts of thermodynamic data can be found empirically as a function of temperature and pressure. From this data a function such as the Gibbs free energy can be formulated. This free energy is an example of a *thermodynamic potential.* Other examples are the Helmholtz free energy, the entropy, and the enthalpy. The thermodynamic properties can be found by taking appropriate derivatives of the thermodynamic potentials. The difference between the various potentials is that they are functions of different variables. For example, the Helmholtz free energy is a function of temperature and volume, and the enthalpy is a function of pressure and entropy. In practice, the potential used depends on which variables can be controlled experimentally or which variables allow a theoretical calculation to be done easily. Typically the temperature and pressure are easily controlled for fluids. The Gibbs free energy of a substance is then computed and tabulated from measurements of the volume, specific heat at constant pressure, coefficient of thermal expansion, and the isothermal compressibility. Frequently, an empirical formula can be formulated for the free energy. For water, tables of such data are called steam tables. In the program FLUIDS we use a formula for the Helmholtz free energy of water derived from experimental data.[3] In addition, we use a theoretical model called the *van der Waals* model.

It is convenient to express the macroscopic properties of a substance so that the properties are independent of its size. Intensive variables such as the pressure or temperature do not depend on the size of the system, while extensive variables such as the volume or internal energy are proportional to the size of the system. We can represent the size of a system by its mass, number of molecules, or volume. We will use lower-case letters for extensive variables that have been converted to intensive quantities by dividing by the mass, number of particles, or the volume. The word "specific" is used to denote an extensive quantity that has been converted to an intensive quantity. Hence the specific energy u can be defined as $u = U/N$, where U is the total internal energy, and N is the number of molecules.

The Helmholtz free energy is defined as $F = U - TS$, where T is the absolute temperature, and S is the entropy. The natural variables for the specific Helmholtz free energy f are the specific volume, $v = 1/\rho$, and T or $\beta = 1/k_B T$. Here ρ is the density, which could be the number or mass density. Given $f(v, T)$, all other thermodynamic quantities can be computed. Some examples are

$$P = -\frac{\partial f}{\partial v} \tag{2.1}$$

$$u = \frac{\partial}{\partial \beta}(\beta f) \tag{2.2}$$

$$s = -\frac{\partial f}{\partial T} \tag{2.3}$$

$$c_V \equiv T\left(\frac{\partial s}{\partial T}\right)_V = -T\frac{\partial^2 f}{\partial T^2} \tag{2.4}$$

$$\kappa_T \equiv -\frac{1}{v}\left(\frac{\partial v}{\partial P}\right)_T = \frac{1}{v}\left(\frac{\partial^2 f}{\partial v^2}\right)^{-1} \tag{2.5}$$

$$\alpha_P \equiv -\frac{1}{v}\left(\frac{\partial v}{\partial T}\right)_P = \frac{1}{v}\left(\frac{\partial^2 f}{\partial v^2}\right)^{-1}\left(\frac{\partial^2 f}{\partial v \partial T}\right) \tag{2.6}$$

$$c_V - c_P = -T\left(\frac{\partial^2 f}{\partial T \partial v}\right)^2\left(\frac{\partial^2 f}{\partial v^2}\right)^{-1}, \tag{2.7}$$

where P is the pressure, c_V is the specific heat at constant volume, c_P is the specific heat at constant pressure, α_P is the coefficient of thermal expansion, and κ_T is the isothermal compressibility.

In the above equations all partial derivatives of the free energy with respect to v assume T is held fixed and vice versa. Even though other quantities assume P is held fixed, they always can be expressed in terms of partial derivatives of f with T or v held fixed. To do so, we consider any two functions $w = w(x, z)$ and $z = z(x, y)$. Because z is a function of x and y, we can also consider w to be a function of x and y; i.e., $w = w(x, z(x, y)) = w(x, y)$. Then the following general mathematical relations hold:

$$\left(\frac{\partial x}{\partial y}\right)_z = \left(\frac{\partial y}{\partial x}\right)_z^{-1} \tag{2.8}$$

$$-1 = \left(\frac{\partial x}{\partial y}\right)_z\left(\frac{\partial y}{\partial z}\right)_x\left(\frac{\partial z}{\partial x}\right)_y \tag{2.9}$$

$$\left(\frac{\partial w}{\partial x}\right)_y = \left(\frac{\partial w}{\partial x}\right)_z + \left(\frac{\partial w}{\partial z}\right)_x\left(\frac{\partial z}{\partial x}\right)_y. \tag{2.10}$$

For example, a thermodynamic function $Y(P)$ can be determined from partial derivatives of Y and P with T or v held constant. Using Eqs. 2.10 and 2.9 we obtain

$$\left(\frac{\partial Y}{\partial T}\right)_P = \left(\frac{\partial Y}{\partial T}\right)_v - \left(\frac{\partial Y}{\partial v}\right)_T\left(\frac{\partial P}{\partial T}\right)_v\left(\frac{\partial P}{\partial v}\right)_T^{-1}. \tag{2.11}$$

If the variation in the temperature and volume is known, it is straightforward although perhaps tedious to calculate all other quantities from the Helmholtz free energy and its derivatives. If the pressure is held fixed (isobaric process) or the entropy is fixed (adiabatic process), it is not so easy. In principle, we can write for an adiabatic process $ds = 0 = (\partial s/\partial v)\, dv + (\partial s/\partial T)\, dT$ and numerically integrate this equation to find the variation of v with T. However, this integration requires a great deal of numerical effort, particularly near phase transitions where the partial derivatives become very large, and thus finite difference approximations to dv and dT must be set very small. If a simple approximate formula for the free energy exists, we might be able to solve for any quantity as a function of any two others. In the next section we describe a model that goes a long way toward achieving this goal.

2.2.3 The van der Waals Model

The van der Waals theory is a simplified model of the gaseous and liquid states of matter that is able to give quantitative data in approximate agreement with empirical data for real fluids made of simple molecules such as argon, water, or oxygen. It is defined by the two equations of state:[4]

$$(P + a\frac{N^2}{V})(V - Nb) = Nk_BT \tag{2.12}$$

$$U = \frac{3}{2}Nk_BT - \frac{aN^2}{V}, \tag{2.13}$$

where a is an empirical constant that accounts for the attraction between molecules due to the dipole-dipole interaction, and b is a constant that accounts for the hard core repulsion at very small intermolecular separations.

One of the triumphs of the van der Waals theory is that it predicts the law of corresponding states, which states that the equations of state of all fluids are identical if the equation of state is rewritten in terms of the ratio of each thermodynamic variable to its critical value. That is, one equation of state rewritten with the variables P, v, and T scaled by their critical values can approximately account for the data for many different kinds of fluids.

We will now convert the van der Waals equations of state to a form where the law of corresponding states is manifest. Recall that at the critical pressure P_c and critical temperature T_c, the first and second derivatives of P with respect to V vanish. Using Eq. 2.12 we find that the critical values of the thermodynamic variables in terms of a and b are given by

$$V_c = 3Nb \tag{2.14}$$

$$P_c = \frac{a}{27b^2} \tag{2.15}$$

$$k_BT_c = \frac{8a}{27b} \tag{2.16}$$

$$U_c \equiv \frac{3}{2}Nk_BT_c - \frac{aN^2}{V_c} = \frac{a}{9b}N. \tag{2.17}$$

If we define all quantities in terms of these critical quantities, i.e., $t = T/T_c$, $p = P/P_c$, $v = V/V_c$, and $u = U/U_c$, the equations of state become

$$\left(p + \frac{3}{v^2}\right)\left(v - \frac{1}{3}\right) = \frac{8}{3}t \tag{2.18}$$

$$u = 4t - \frac{3}{v}. \tag{2.19}$$

Note that in this context lower-case quantities denote thermodynamic variables divided by their critical values. The fact that Eqs. 2.18 and 2.19 do not depend on the constants a and b is an example of the law of corresponding states.

We now can construct a free energy f such that $p = -\partial f/\partial v$ and $u = -\partial(f/t)/\partial(1/t)$. The result is

$$f = -\frac{3}{v} - \frac{8t}{3}\left(1 + \ln(v - \frac{1}{3}) + \frac{3}{2}\ln t\right). \tag{2.20}$$

Using Eq. 2.20, we find that the entropy is

$$s = -\frac{\partial f}{\partial t} = 4 + \frac{8}{3}\left(1 + \ln(v - \frac{1}{3}) + \frac{3}{2}\ln t\right). \tag{2.21}$$

One desirable feature of Eqs. 2.18–2.21 is that we can solve for v or t from any of the pairs of variables (p, t), (p, v), (s, t), (s, v), (u, t), and (u, v). This fact makes it computationally feasible to investigate quickly a large number of thermodynamic processes.

The free energy given in Eq. 2.20 is not correct over all ranges of t and v. For certain values of (t, v), the compressibility can be negative, an unphysical result. In addition, as can be seen in Figure 2.2, there are values of the pressure, for which there are three possible values for the volume. This situation is an indication of a phase transition between a liquid and a gas. Below the critical pressure and temperature, it is possible for liquid and gas to coexist. When this happens, both phases have the same pressure and temperature, but different volumes. They also must have the same Gibbs free energy per particle G/N, because G/N equals the chemical potential for each phase. Two materials, such as a liquid and its vapor, can be in equilibrium only if their chemical potentials are equal. Otherwise,

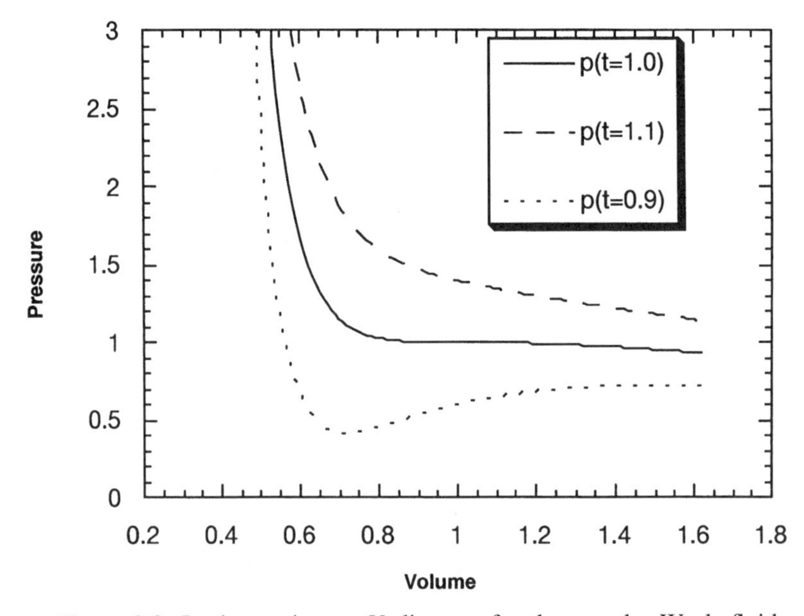

Figure 2.2: Isotherms in a p-V diagram for the van der Waals fluid.

molecules of the higher chemical potential would diffuse into the region of the lower chemical potential.

If we equate the Gibbs free energy of the two phases, we obtain the so-called Maxwell equal area construction illustrated in Figure 2.3. The correct pressure is determined by finding the pressure such that the areas between the curve and a horizontal line are equal ($A_1 = A_2$ in Fig. 2.3). To save computing time, the calculation of these pressures for a number of isotherms, and the corresponding liquid and gas volumes, v_l and v_g, respectively, has been performed in advance and is stored in a coexistence table for use by the program FLUIDS. The free energy and any derivative is calculated by first checking to see if $t < 1$. If so, the corresponding values of v_g and v_l are found in the table; if $v_l < v < v_g$, the free energy is given by

$$x = \frac{v - v_l}{v_g - v_l} \tag{2.22}$$

$$f(t, v) = x f(t, v_g) + (1 - x) f(t, v_l), \tag{2.23}$$

instead of Eq. 2.20. Similar calculations are done when other quantities are given. For example, if p and v are given and p is less than unity, the program checks the coexistence table to see if $v_l < v < v_g$ for the given p. If so, the temperature is found from the table instead of Eq. 2.18. If p and t are given, the program first solves for v by solving a cubic equation. If there is only one real root, that is the correct volume. If there are three real roots, the program uses the table to find out if p is greater than or less than the coexistence pressure at t. If it is greater, i.e.,

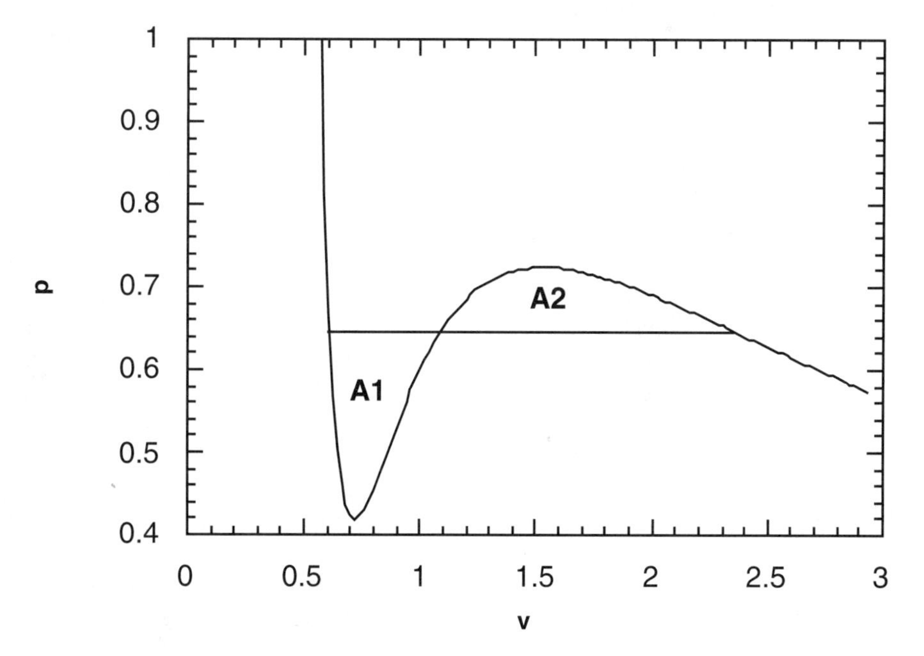

Figure 2.3: Maxwell equal area construction. The curve corresponds to a temperature $t = 0.9$. The pressure at two-phase coexistence is $p \approx 0.65$, with the liquid volume $v_1 \approx 0.60$ and the gas volume $v_g \approx 2.35$.

the system is a liquid, then the correct volume is the smallest root; otherwise it is the largest root. The coexistence table also includes the entropy and energy for the liquid and gas. Given (s, t) or (u, t) with $t < 1$, the program finds the coexistence values in the table. If the given values s or u, respectively, are between the liquid and gas values, the program interpolates to find the fraction x of liquid, and uses Eq. 2.22 to find v. It is too difficult to determine T from (s, v) or (u, v) in the coexistence region, and hence the program does not allow the user to draw a path in the s-v or u-v diagrams.

2.2.4 Water

The van der Waals model does surprising well in describing the qualitative properties of many fluids. For detailed quantitative data over a full range of macroscopic variables, and for precise values near the critical point, a better equation of state is needed. One of the most important fluids is water. Because of its importance we have incorporated an empirical formula for the free energy of water into the program FLUIDS.

The Helmholtz free energy used in the program FLUIDS for water is an empirical formula.[3] This formula uses over 50 empirically determined constants to obtain a free energy that is accurate for most values of temperature and volume where water exists as a liquid or a gas. Keenan et al.[3] also provide the coexistence table for liquid water and its vapor. (This table also is found in Callen's text.[4]) The program reads in the coexistence table, and uses Eq. 2.23 in the coexistence region. The empirical formula for $f(v, T)$ and its derivatives are too complicated to solve for v or T. For example, if p and v are given, it is very difficult to solve numerically for T from $p = -\partial f/\partial v$. Thus, the FLUIDS program allows the user to draw only in the v-T diagram; it then can produce plots in the other phase diagrams.

2.3 Procedure for Running the Program FLUIDS

The FLUIDS program is designed so that the user can explore the representations of thermodynamic paths in different kinds of phase diagrams. The program also can help you obtain some feel for the order of magnitude of various thermodynamic quantities. To run the program follow these steps:

1. After you start up the program, data will be read in for the coexistence tables. When that is finished, a brief description of the program appears. After clicking the mouse or any key input, an input screen will appear.

2. From the input screen choose the type of fluid, water or van der Waals, and the phase diagrams you wish to see. The program can show four phase diagrams at once. The possible phase diagram choices are

    a. v-T.
    b. P-T.

c. P-v.
d. s-v.
e. s-T.
f. u-v.
g. u-T.

After you make your choices, four windows will appear, each containing a phase diagram. The P-T diagram shows the coexistence curve, and the other diagrams show the boundary between the coexistence region and the single phase regions.

3. Once the phase diagrams are shown, you can draw a straight-line path in a window by holding the mouse button down at the first point of the line segment and letting go at the second point. For the van der Waals model you may draw the path in any plot except the s-v and u-v diagrams. For water you may draw a path only in the v-T diagram. After you draw a path, the paths in the other diagrams are calculated and drawn. For water this calculation may take a few seconds on slower computers.

4. You may continue to draw as many paths as you wish. Four colors are used to help you distinguish one path from another.

5. The **F2-Clear** hot key will erase all the paths and redraw the phase diagrams. The **F3-New** hot key will show you the input screen so that you can change your choice of fluid or which phase diagrams to see. The **F4-See Data** hot key displays numerical values for the following thermodynamic quantities for the last point plotted:

   - Pressure
   - Temperature
   - Specific Volume
   - Specific Energy
   - Specific Entropy
   - Specific Helmholtz free energy
   - Specific Heat at Constant Volume
   - Isothermal Compressibility
   - $c_p - c_v$
   - Coefficient of Thermal Expansion

   You can use this feature to compile data on many quantities by clicking and letting go at the same point. The **F10-Menu** hot key will return you to the main menu.

6. Choose **Exit Program** from the **File** heading of the main menu to exit the program.

2.4 Exercises

2.1 **Use of Partial Derivatives**
Derive Eqs. 2.5–2.7 using Eqs. 2.8–2.10.

2.2 **Water and van der Waals Phase Diagrams**
Use the program to compare the phase diagrams for water and the van der Waals fluid. Can you detect any qualitative features that are different? If so, explain how the properties of the intermolecular forces of water might lead to these differences.

2.3 **Gas and Liquid Properties**
Choose three points on any phase diagram such that one point is a liquid, one point is a gas, and one point is above the critical point. Record the thermodynamic properties of each of the three points and discuss their relative sizes given what you know about gases and liquids.

2.4 **Specific Heat of Water**
Choose five points on the V-T diagram for water and another five points for the van der Waals model. Be sure to choose at least one point in each of the gas, liquid, and two phase coexistence regions. Record the specific heat at constant volume. Discuss the values you obtain paying particular attention to why the van der Waals model does not account for the variation found in water.

2.5 **Claussius-Clapeyron Equation**
The Claussius-Clapeyron equation relates the derivative dP/dT on the coexistence curve to the latent heat and volume change on crossing the coexistence curve. The equation is $dP/dT = \Delta s/\Delta v$, where $\Delta s = s_g - s_l$ and $\Delta v = v_g - v_l$. Use the program to estimate dP/dT by finding P and T for two nearby points on the coexistence curve for the van der Waals fluid. Use the s-v plot to determine Δs and Δv as one crosses the coexistence curve between these two points. Does the Claussius-Clapeyron equation hold? Repeat this exercise at another point on the coexistence curve.

2.6 **Positivity of the Specific Heat**
Using the van der Waals fluid, show the u-T and P-v or v-T diagrams. Now draw a constant volume path. Is the slope of the path in the u-T diagram always positive? Remember this slope equals c_V. Repeat for a few other paths. Does the slope change much in different parts of the phase diagram? Discuss your results. Compare with u-T and v-T diagrams for water. Now use a similar approach for c_P. Where is c_P greatest? What is its value inside the two-phase region? Explain in physical terms what you find.

2.7 **Positivity of Compressibility**
Recall the definition $\kappa_T = -(1/v)(\partial v/\partial P)_T$. Show that $\kappa_T > 0$ by drawing isotherms and looking at the slopes of the paths in the P-v diagram. For the van der Waals fluid, show that κ_s, the adiabatic compressibility, also is non-negative. Which do you expect to be greater, κ_T or κ_s? What do

you find from the program? What value does κ_T take on in the two-phase region? Explain why κ_T must be non-negative to maintain mechanical stability.

2.8 **First-Order Phase Transitions**
Collect thermodynamic data for the van der Waals fluid from an isobaric path containing ten points (five gas and five liquid) in the P-T plane that crosses the coexistence curve. For each of the points record the free energy, volume, energy, entropy, isothermal compressibility, coefficient of thermal expansion, the specific heat at constant volume, and $c_P - c_V$. Which of these quantities are discontinuous as you pass through the phase boundary? Which have discontinuous first derivatives with respect to temperature? For some of these quantities you can see their behavior immediately by drawing an isobaric path in the P-T plane and seeing what happens in the other phase diagrams.

2.9 **Continuous Phase Transitions**
Collect thermodynamic data for the van der Waals fluid from an isobaric path containing five points in the P-T plane that crosses the critical point. For each of the points record the free energy, volume, energy, entropy, isothermal compressibility, coefficient of thermal expansion, the specific heat at constant volume, and $c_P - c_V$. Do any of these quantities diverge at the critical point? For some of these quantities you can see their behavior immediately by drawing an isobaric path in the P-T plane and seeing what happens in the other phase diagrams.

2.10 **Properties of Water**
Find the point on the v-T diagram that is at atmospheric pressure and room temperature (about 22^oC). Record the thermodynamic properties. Discuss their numerical values given what you know about water. Repeat this exercise for steam at atmospheric pressure.

2.11 **Carnot Cycles**
Draw a rectangular path in the gas part of the s-T diagram for the van der Waals fluid. This path corresponds to a Carnot cycle. Record the thermodynamic data at the corners of the cycle. Discuss the shape of the cycle in the P-v diagram and calculate the work done by the cycle, the heat taken up at the hot end of the cycle, and the efficiency. Now draw another rectangular path that includes the two-phase region. Compare the work done and the efficiency of this second cycle with that of the first.

2.12 **Work**
The work done by a system is given by $\Delta w = du - \Delta q = du - Tds$, where the latter equality holds for reversible processes. Use the program with the van der Waals fluid to confirm that the work done is given by the change in the free energy for isothermic processes, or the change in enthalpy, $(u + Pv)$, for isobaric processes, and equals the change in the Gibbs free energy for processes at constant temperature and pressure.

2.13 **Two Phase Coexistence**
Consider a closed container consisting of a liquid in equilibrium with its vapor. The container is heated up so that the temperature inside increases.

Since the volume of the container is fixed, this process corresponds to a horizontal path in the v−T phase diagram. Draw three horizontal paths: $v < v_2$, $v = v_c$, and $v > v_c$. Each path should begin inside the two phase region and end outside of it. Describe the corresponding paths in the p−T phase diagram. Describe what happens in the container for each of the three paths.

2.5 Program Modifications

The most significant program modification would be to introduce another Helmholtz free energy for a fluid. There are other equations of state more sophisticated than the van der Waals model, but they do not provide much more insight.

A simple modification would be to add other thermodynamic quantities such as the Gibbs free energy or the enthalpy to the data calculated by the FLUIDS program. This modification can be done by writing a quantity in terms of other quantities already calculated and/or the appropriate derivatives of the Helmholtz free energy, and then adding these calculations to Procedure **ShowData**.

Another modification would be to allow the user to draw paths in all the phase diagrams. For example, let us say we drew a path in the P-T diagram for water, with the first point corresponding to a pressure and temperature (P_1, T_1). First we would need a previously stored table of P as a function of v and T calculated from the free energy using $P = -\partial f/\partial v$. Then we could either interpolate to find the value v_1 that corresponds to (P_1, T_1) or we could calculate it using

$$dp = \frac{\partial p}{\partial v} dv + \frac{\partial p}{\partial T} dT , \quad (2.24)$$

where $dp = P - P_1$, $dv = v - v_1$, and $dT = T - T_1$. Here P and T would be the table values closest to (P_1, T_1). The partial derivatives in Eq. 2.24 are simply second derivatives of the free energy, for which there are procedures already in the program. Then to find the values for v for the rest of the path we could use Eq. 2.24 with small values of dp and dT to march along the curve. The difficulty with this procedure is that the size of dp and dT necessary to produce accurate results depends on where you are in the P-T diagram. In particular you must be extremely accurate near the coexistence curve where the partial derivatives in Eq. 2.24 can diverge.

References

1. Sears, F., Salinger, G. *An Introduction to Thermodynamics, the Kinetic Theory of Gases, and Statistical Mechanics.* 3rd ed. Reading, MA: Addison-Wesley, 1975.

2. Kittel, C., Kroemer, H. *Thermal Physics.* 2nd ed. San Francisco: W.H. Freeman, 1980.

3. Keenan, J.H., Keyes, F.G., Hill, P.G., Moore, J.G. *Steam Tables.* New York: Wiley-Interscience, 1978.

4. Callen, H.B. *Thermodynamics and an Introduction to Thermostatistics.* 2nd ed. New York: John Wiley and Sons, 1985.

3

Engines

L. B. Spornick

3.1 Introduction

The program ENGINES provides an introduction to the thermodynamic processes involved in converting heat into work. Users can design and study engines or interpret the results of animations of a Diesel, an Otto, and a Wankel engine.

3.2 Engines

An engine is a mechanical device that works in a cycle to convert heat into work. The engine repeats a series of processes, and at the end of the series, the engine returns to its initial condition.

There are basically two types of engines—internal combustion and external combustion. In an internal combustion engine, heat is supplied by the combustion of gases inside the engine. The gasoline engine and the Diesel engine are examples of internal combustion engines. In an external combustion engine, heat is supplied by an outside source. The Stirling engine and the steam engine are examples of external combustion engines.

3.2.1 Reversible Engines

An engine's cycle consists of a series of processes. These processes can be either reversible or irreversible. During a reversible process, the engine's thermodynamic properties are well defined. The engine is always in thermal equilibrium. Engines whose processes are reversible are called reversible engines. In contrast, during an irreversible process, the engine's thermodynamic properties may not be defined. Engines whose processes are irreversible are called irreversible engines.

All real engines are irreversible. To describe their cycles in terms of thermodynamic processes, approximations have to be made, including representing an irreversible process by a reversible one and using ideal gases in the engine.

The equation of state for a system describes the relationship between the system's thermodynamics coordinates (temperature, volume, pressure, etc.) when the system is in thermal equilibrium. The equation of state for an ideal gas is

$$PV = nRT, \tag{3.1}$$

where P is the pressure, V is the volume, n is the number of moles of the gas (which is assumed to be 1.), R is the gas constant, T is the temperature in Kelvins, and the entropy (S) is

$$S = R(\frac{3}{2} + \ln(eV(T/\lambda)^{\frac{3}{2}})), \tag{3.2}$$

where e is the base of the natural logarithm (ln), $\lambda = h/\sqrt{2\pi M k_B}$, h = Planck's constant, M is the mass of a molecule and k_B is the Boltzmann's constant. This entropy term represents only the contribution from the translational motion of the molecules in the gas. For multi-atomic molecules, there are additional contributions from the molecular vibrations and rotations.

The reversible thermodynamic processes for an ideal gas are *adiabatic* (constant entropy, no heat exchanged), *isobaric* (constant pressure), *isochoric* (constant volume, also called *isovolumic*), and *isothermal* (constant temperature).

An adiabatic process is one in which no heat is exchanged between the engine and its surroundings. The engine is thermally isolated and its entropy (S) is constant. For an ideal gas, the relationships between the initial and final temperatures, volumes and pressures are

$$T_i V_i^{\gamma-1} = T_f V_f^{\gamma-1} \tag{3.3}$$

$$P_i V_i^{\gamma} = P_f V_f^{\gamma} \tag{3.4}$$

$$T_i P_i^{-(\gamma-1)/\gamma} = T_f P_f^{-(\gamma-1)/\gamma}, \tag{3.5}$$

where γ is the ratio of the specific heat at constant pressure (C_P) to the specific heat at constant volume (C_V). The work done by the engine is $(P_i V_i - P_f V_f)/(\gamma - 1)$ and the change in the engine's internal energy is $C_V(T_f - T_i)$.

An isobaric process is one in which the pressure is kept constant. For an ideal gas, the relationships between the initial and final temperatures, volumes, and entropies are

$$T_i/V_i = T_f/V_f \tag{3.6}$$

and

$$S_f = S_i + C_P \ln(T_f/T_i). \tag{3.7}$$

The work done by the engine is $P_f(V_f - V_i)$, the heat that is absorbed is $C_P(T_f - T_i)$ and the change in the engine's internal energy is $C_V(T_f - T_i)$.

An isochoric process is one in which the volume is kept constant. For an ideal gas, the relationships between the initial and final temperatures, pressures, and entropies are

$$T_i/P_i = T_f/P_f \tag{3.8}$$

and

$$S_f = S_i + C_V \ln(T_f/T_i). \tag{3.9}$$

No work is done by the engine, the heat that is absorbed is $C_V(T_f - T_i)$, and the change in the engine's internal energy is $C_V(T_f - T_i)$.

An isothermal process is one in which the temperature is kept constant. For an ideal gas, the relationships between the initial and final volumes, pressures, and entropies are

$$V_i P_i = V_f P_f \tag{3.10}$$

and

$$S_f = S_i + R \ln(V_f/V_i). \tag{3.11}$$

The work is done by the engine is $RT \ln(V_f/V_i)$, the heat that is absorbed is $T(S_f - S_i)$ and there is no change in the engine's internal energy.

These relationships are summarized in Tables 3.1 and 3.2.

Table 3.1: Equations for governing the thermodynamic parameters for adiabatic, isobaric, isochoric, and isothermal processes for an ideal gas.*

Process	Final temperature specified	Final volume specified	Final pressure specified
Adiabatic	$V_f = V_i(T_i/T_f)^{1/(\gamma-1)}$ $P_f = R T_f/V_f$ $S_f = S_i$	$T_f = T_i(V_i/V_f)^{\gamma-1}$ $P_f = R T_f/V_f$ $S_f = S_i$	$V_f = V_i(P_i/P_f)^{1/\gamma}$ $T_f = V_f P_f/R$ $S_f = S_i$
Isobaric	$V_f = V_i T_f/T_i$ $P_f = P_i$ $S_f = S_i + C_P \ln T_f/T_i$	$T_f = T_i V_f/V_i$ $P_f = P_i$ $S_f = S_i + C_P \ln V_f/V_i$	
Isochoric	$P_f = P_i T_f/T_i$ $V_f = V_i$ $S_f = S_i + C_V \ln T_f/T_i$		$T_f = T_i P_f/P_i$ $V_f = V_i$ $S_f = S_i + C_V \ln P_f/P_i$
Isothermal		$T_f = T_i$ $P_f = P_i V_f/V_i$ $S_f = S_i + R \ln V_f/V_i$	$V_f = V_i P_f/P_i$ $T_f = T_i$ $S_f = S_i + R \ln P_f/P_i$

*We use standard notation; e.g., S is the entropy, C_V is the specific heat at constant volume, C_P is the specific heat at constant pressure, and $\gamma = C_P/C_V$.

Table 3.2: Work done, heat absorbed and change in the internal energy for adiabatic, isobaric, isochoric, and isothermal processes for an ideal gas of 1 mole.

Process	Work done to the surroundings	Heat absorbed from the surroundings	Change in gas's internal energy
Adiabatic	$(P_i V_i - P_f V_f)/(\gamma - 1)$	None	$C_V(T_f - T_i)$
Isobaric	$P_f (V_f - V_i)$	$C_P (T_f - T_i)$	$C_V (T_f - T_i)$
Isochoric	None	$C_V (T_f - T_i)$	$C_V (T_f - T_i)$
Isothermal	$RT_f \ln V_f/V_i$	$T_f (S_f - S_i)$	None

3.2.2 Engines and the Second Law of Thermodynamics

Rudolf Clausius's statement of the second law of thermodynamics is

It is impossible to build an engine which, working in a cycle, only transfers heat from a colder body to a hotter body.

The important phrase is *"working in a cycle"* because it is possible to construct a non-cyclic process in which heat is extracted from a colder body and absorbed by a hotter body and not violate the second law.

Consider the following series of processes, which uses a gas enclosed in a cylinder with a piston on top. Initially, the gas is in thermal contact with a heat source at temperature T_1, where $T_1 < T_2$ (Fig. 3.1).

- Process 1: The piston rises, increasing the gas's volume and extracting heat from the heat source T_1. This process is isothermal (Fig. 3.2).

- Process 2: The gas is thermally isolated and compressed until its temperature is T_2. This process is adiabatic (Fig. 3.3).

- Process 3: The gas is now in thermal contact with a heat source at temperature T_2 and compressed further. Heat is transferred from the gas to the heat source at temperature T_2. This process is isothermal (Fig. 3.4).

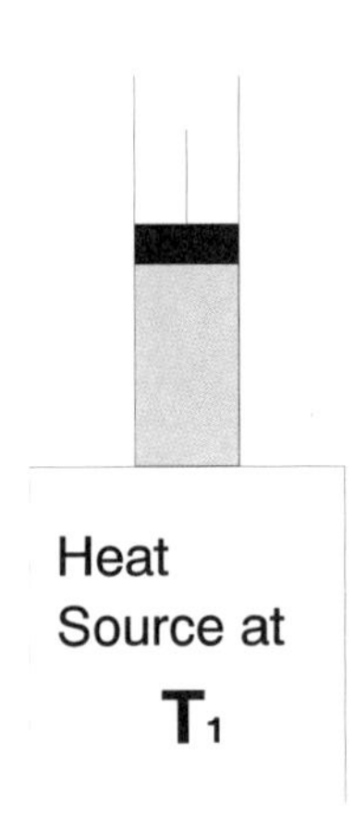

Figure 3.1: Gas in thermal contact with a heat source at T_1.

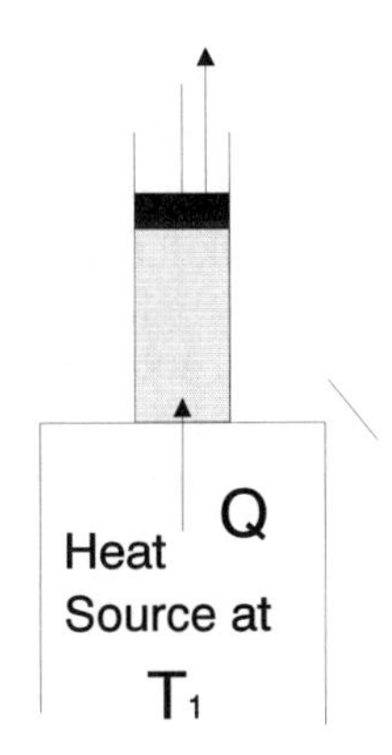

Figure 3.2: Isothermal expansion.

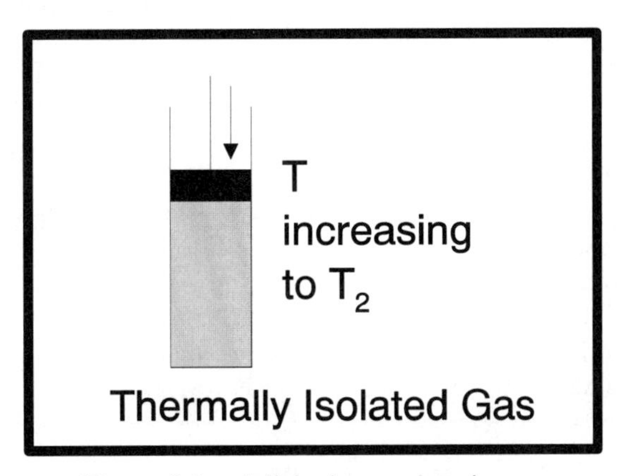

Figure 3.3: Adiabatic compression.

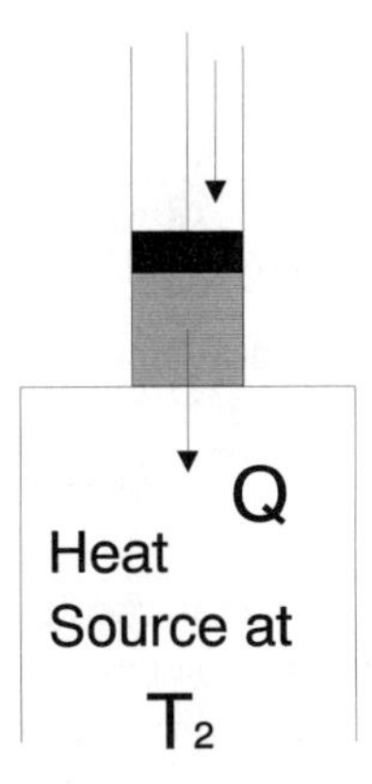

Figure 3.4: Isothermal compression.

In this example, heat is transferred from a colder body at T_1 to a hotter body at T_2. But it is not a violation of the second law. This series of processes is not cyclic because the gas's final state (T_2) is not the same as the initial state (T_1).

Lord Kelvin proposed another version of the second law:

It is impossible to build a device which, working in a cycle, produces no effect other than the extraction of heat from a reservoir and the performance of an equal amount of mechanical work.

It means that an engine cannot extract heat from one source and convert it all to usable work. An engine must have access to at least two heat reservoirs: a source which supplies heat to the engine and a sink which receives the excess heat (i.e., the energy that is not converted to work).

Although the Clausius and the Kelvin forms of the second law look different, they are equivalent.

The second law of thermodynamics can be expressed as

$$\Delta S_{total} \geq 0, \tag{3.12}$$

where ΔS_{total} is, for a thermodynamic process, the change in the total entropy of the system. Entropy is an extensive quantity and is a function of the size of the system. Gibbs' law of partial entropies states that the total entropy of an isolated system equals the sum of the entropies of the parts of the system. For an engine, the total entropy is the sum of the engine's entropy and the entropy of the rest of the universe.

Because of the second law of thermodynamics, the change in total entropy of the system is greater than or equal to zero. The change in the total entropy equals zero only for reversible processes. An engine operates in a cycle, returning to its initial conditions, so $\Delta S_{engine} = 0$. The rest of the universe does not return to its original conditions, so the change in entropy of the rest of the universe, $\Delta S_u \geq 0$. ($\Delta S_u = 0$, only if the engine is reversible.)

3.2.3 Engine Efficiency

The *efficiency* of an engine is defined as the ratio of the amount of net work done by the engine, W, to the amount of heat that is absorbed from the heat source, Q_h:

$$\text{Efficiency} = \frac{W}{Q_h}. \tag{3.13}$$

The first law of thermodynamics can be written as

$$\Delta U = Q - W, \tag{3.14}$$

where ΔU is the change in the engine's internal energy and $Q = Q_h - |Q_c|$, with Q_c = the heat that is discharged to the sink. At the end of a cycle, the internal energy of an engine returns to its initial value, and hence $\Delta U = 0$ and $W = Q_h - |Q_c|$. So the *efficiency* of an engine is given by

$$\text{Efficiency} = 1 - \frac{|Q_c|}{Q_h}. \tag{3.15}$$

The area enclosed by a plot of P versus V equals W, and the area enclosed by a plot of T versus S equals Q.

3.2.4 The Carnot Engine

In 1824, the French physicist Sadi Carnot published a paper in which he defined a reversible ideal gas engine. The engine's cycle consists of the following four processes:

- Process 1: An Isothermal Compression
  The engine's temperature remains constant (T_1), while its volume decreases from V_1 to V_2, its pressure increases from P_1 to P_2, and its entropy decreases from S_1 to S_2 (Fig. 3.5). From Table 3.1, given V_2, $P_2 = P_1(V_1/V_2)$ and $S_2 = S_1 + R\ln(V_2/V_1)$. From Table 3.2, the heat that is expelled by the engine equals $T_1(S_2 - S_1)$, and the work done on the engine is $RT_1 \ln(V_2/V_1)$. Since the internal energy of an ideal gas is only a function of the temperature, the internal energy is unchanged.

- Process 2: An Adiabatic Compression
  The engine's temperature increases from T_1 to T_2, its volume decreases from V_2 to V_3, its pressure increases from P_2 to P_3, and its entropy remains constant (S_2) (Fig. 3.6). Given T_2, $V_3 = V_2(T_1/T_2)^{1/(\gamma-1)}$ and $P_3 = RT_2/V_3$. No heat is exchanged between the engine and its surroundings. The work done to the engine is $(P_2V_2 - P_3V_3)/(\gamma - 1)$ and the change to the internal energy is $C_V(T_2 - T_1)$.

- Process 3: An Isothermal Expansion
  The engine's temperature remains constant (T_2), while its volume increases from V_3 to V_4, its pressure decreases from P_3 to P_4, and its entropy returns to S_1 (Fig. 3.7). V_4 is determined by the requirement that the system return to its initial conditions (T_1, V_1, P_1, S_1) after an adiabatic expansion (process 4): $V_4 = V_1(T_1/T_2)^{1/(\gamma-1)}$, $P_4 = P_3(V_3/V_4)$. The heat that is absorbed by the engine at T_2 is $T_2(S_1 - S_2)$ and the work done by the engine is $RT_2 \ln(V_4/V_3)$. The internal energy is unchanged.

- Process 4: An Adiabatic Expansion Back to the Initial State
  The engine's temperature decreases from T_2 to T_1, its volume increases from V_4 to V_1, its pressure decreases from P_4 to P_1, and its entropy remains constant

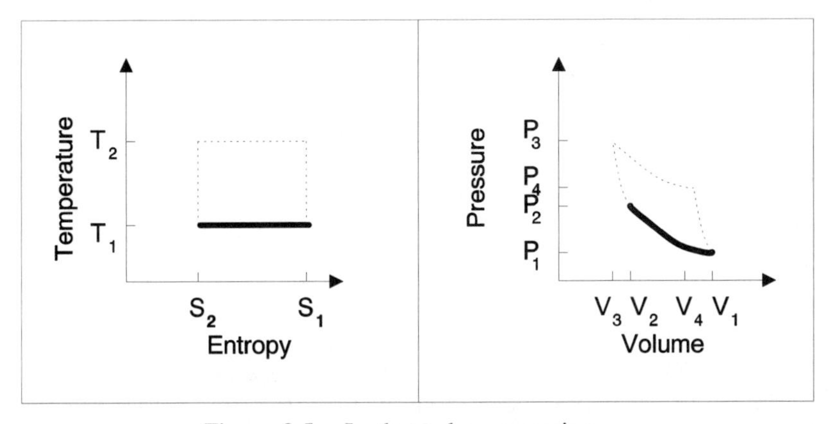

Figure 3.5: Isothermal compression.

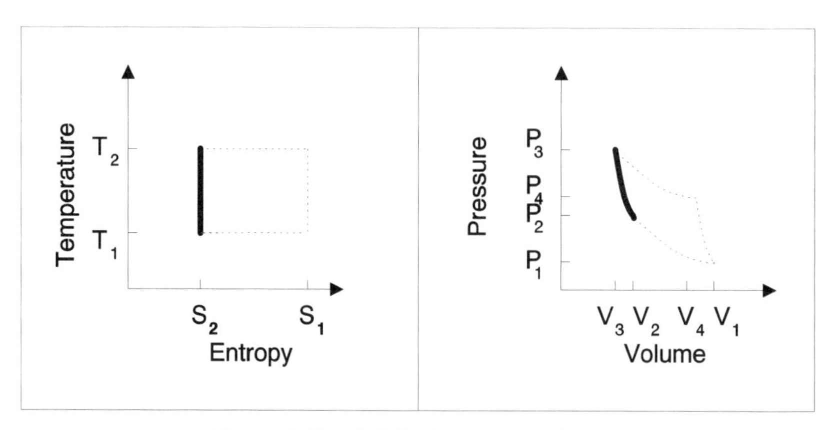

Figure 3.6: Adiabatic compression.

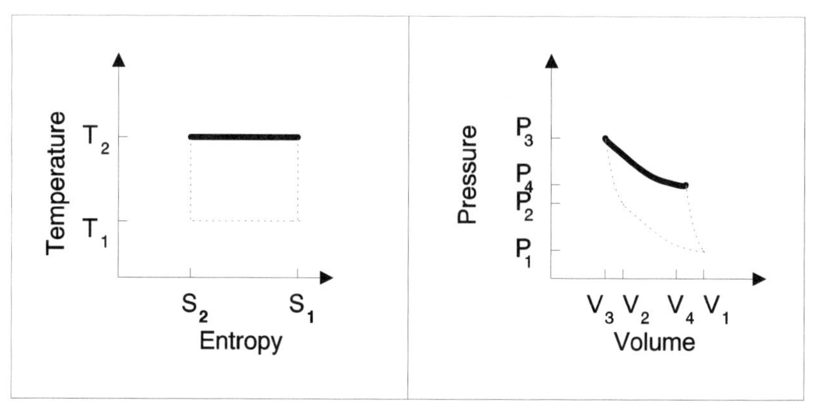

Figure 3.7: Isothermal expansion.

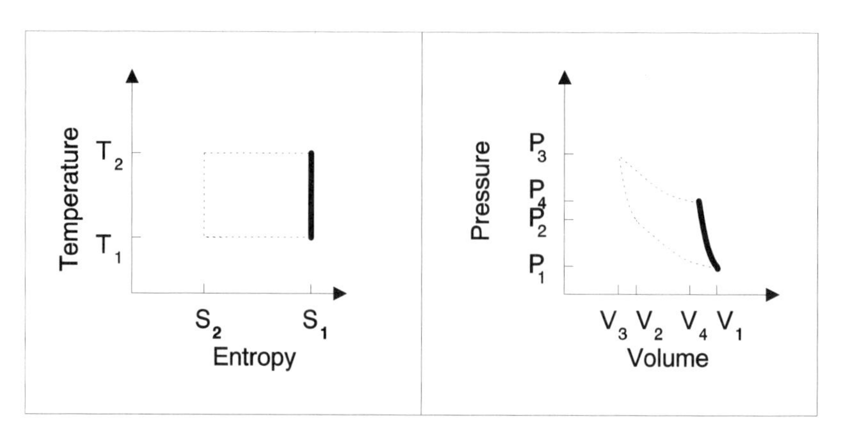

Figure 3.8: Adiabatic expansion.

(S_1) (Fig. 3.8). No heat is exchanged between the engine and its surroundings. The work done by the engine is $(P_4V_4 - P_1V_1)/(\gamma - 1)$ and the change to the internal energy is $C_V(T_1 - T_2)$.

The area enclosed in the T versus S plot equals the total heat absorbed and the area enclosed in the P versus V plot equals the work done by the engine.

Carnot proved that, for a reversible engine, the amount of work produced and the engine's efficiency depends only upon the sources' and sinks' temperatures. Lord Kelvin later used this relationship to devise the absolute temperature scale that bears his name. The efficiency for a Carnot engine is

$$\text{Efficiency} = 1 - \frac{T_c}{T_h}. \tag{3.16}$$

Because the third law of thermodynamics precludes attaining a temperature of absolute zero in a finite number of steps, the efficiency of an engine can never be equal to 1.

3.2.5 The Diesel, Otto, and Wankel Engines

The idealized Diesel, Otto, and Wankel (rotary) engines can be used to describe most automobile engines, even though an automobile engine is an irreversible engine. These idealized engines assume that the working substances are ideal gases, all processes are reversible, and there is no friction. (See Ref. 1, 2, and 5.)

The Diesel engine has six processes:

- Process 1: The *intake stroke* in which air is drawn into the combustion chamber by the motion of the piston. This process is isobaric and isothermal.

- Process 2: The *compression stroke* in which the air in the combustion chamber is compressed by the motion of the piston. The compression continues until the temperature of the air is high enough to ignite the oil that will be sprayed into the chamber in the next process. This process is adiabatic.

- Process 3: The *explosion* in which the oil that is sprayed into the combustion chamber is ignited. The volume of the combustion chamber changes so that the pressure remains constant, and hence the process is isobaric.

- Process 4: The *power stroke* in which the hot gases cause the piston to move. This process is adiabatic.

- Process 5: The *valve exhaust* in which there is a drop in pressure and temperature caused by the quasistatic (reversible) ejection of heat due to the opening of the exhaust valve. This process is isochoric.

- Process 6: The *exhaust stroke* in which the piston moves, pushing out the combustion gases. This process is isobaric and isothermal and cancels Process 1.

The Otto engine represents an idealized gasoline engine and consists of six processes:

- Process 1: The *intake stroke* in which a mixture of gasoline vapor and air is drawn into the combustion chamber by the movement of the piston. The process is isobaric and isothermal.

- Process 2: The *compression stroke* in which the piston moves, compressing the gas mixture. This process is adiabatic.

- Process 3: The *explosion* in which an electric spark ignites the mixture. The piston does not move. This process is isochoric.

- Process 4: The *power stroke* in which the hot gases cause the piston to move. This process is adiabatic.

- Process 5: The *valve exhaust* in which there is a drop in pressure and temperature caused by the quasistatic ejection of heat due to the opening of the exhaust valve. This process is isochoric.

- Process 6: The *exhaust stroke* in which the piston moves, pushing out the combustion gases. This process is isobaric and isothermal and cancels Process 1.

The Wankel engine represent an idealized rotary gasoline engine and also has six processes:

- Process 1: *Intake* in which a mixture of gasoline vapor and air is drawn into the combustion chamber caused by the movement of the rotor. This process is isobaric and isothermal.

- Process 2: *Compression* in which the rotor moves, compressing the gas mixture. This process is adiabatic.

- Process 3: *Explosion* in which an electric spark ignites the mixture. The rotor does not move. This process is isochoric.

- Process 4: *Power* in which the hot gases cause the rotor to move. This process is adiabatic.

- Process 5: *Vent exhaust* in which a drop in pressure and temperature is caused by the quasistatic ejection of heat due to the contact of the combustion gases with the surroundings. This process is isochoric.

- Process 6: *Exhaust* in which the rotor moves, pushing the combustion gases out of the chamber. This process is isobaric and isothermal and cancels Process 1.

Although the Otto engine and the Wankel engine physically differ, the thermodynamic processes are identical.

3.2.6 Refrigerators

A heat engine is a device in which useful work is done to the surroundings by absorbing heat from a source at a high temperature and expelling heat to a sink at a low temperature. A refrigerator is the reverse of a heat engine. Work is done to the refrigerator which extracts heat from a source at a low temperature and expels it to a source at a high temperature. The measurement of the performance of the refrigerator is called the *coefficient of performance,* w:

$$w = \frac{\text{heat extracted from cold reservoir}}{\text{work done to the refrigerator}}. \tag{3.17}$$

3.3 Computational Approach

3.3.1 The Models

The thermodynamic theory of engines in section 3.2 provides the computational approach for the four programs: DIESEL, OTTO, WANKEL, and ENGINE. The first three programs model idealized versions of common automobile engines. The last program allows the user to define an ideal gas engine cycle and study its thermodynamic properties.

3.3.2 The DIESEL, OTTO, and WANKEL Engine Models

DIESEL, OTTO, and WANKEL demonstrate the relations between the movements of idealized physical engines and their thermodynamic properties. DIESEL illustrates an idealized Diesel engine, OTTO illustrates an idealized gasoline engine, and WANKEL illustrates an idealized Wankel or rotary gasoline engine. The initial conditions are $T = 300K$ and $P = 1$ atm for the programs.

3.3.3 The Design Your Own Engine Model

Use ENGINE to design an engine cycle for an ideal gas. The gases allowed are helium, argon, nitrogen, and steam, and the processes allowed are adiabatic, isobaric, isochoric, and isothermal. The engine can be either reversible or irreversible. Irreversibility is simulated by a heat loss during the isobaric, isochoric, and isothermal processes. The user controls the percentage of heat loss.

The program provides a graph of T versus S and P versus V during the cycle and a list of the step number, process type, the final pressure, the final volume, the final temperature, the final entropy, the work done by the gas during the process, the heat that is absorbed by the gas during the process, and the change in the internal energy of the gas. When the cycle is complete, i.e., the gas returns to its initial condition, the program determines if the cycle is an engine (the gas does work on its surroundings) or a refrigerator (the surroundings do work on the gas) and computes either the efficiency of the engine or the coefficient of performance of the refrigerator.

The user defines the gas's initial temperature and volume, then builds the engine by specifying a series of processes and their durations. To select a process, the user presses the appropriate radio button. Once a process is selected, sliders for the temperature, volume, and/or pressure appear. They are used to determine that process's final temperature, volume, pressure, and entropy. The program checks that the values for the temperature, etc., are within the limits specified by the plots. If they are not within these limits, an error message appears on the screen.

3.4 Exercises

An asterisk (*) indicates that the exercise is advanced. ENGINE creates a disk file that contains data that may be helpful in the analysis of the results. This data file is described in section 3.5.

DIESEL Engine, the OTTO Engine, and the WANKEL Engine

3.1 **The Relationship Between the Physical Activities and the Thermodynamic Properties**
Run each of the programs to examine the relationship between the engines' physical processes and the corresponding changes in the thermodynamic properties.

3.2 **The Stirling Engine***
Write a similar program for a Stirling engine.

Design Your Own Engine

3.3 **Relationships Between the Efficiencies, Heat Absorbed, and Work Done by an Otto and a Diesel Engine**

a. Using nitrogen and a reversible Otto engine, for constant T_3 and varying T_2 from 650 K to 850 K, plot the total work produced versus the efficiency and the work produced versus the total heat absorbed. Let T_3 = 1500 K, 1600 K, 1700 K, 1800 K, 1900 K, and 2000 K.

b. Repeat, using a reversible Diesel engine.
T_2 is the temperature of the gas at the end of the adiabatic compression. T_3 is, for the Otto engine, the temperature of the gas at the end of the first isochoric process, and, for the Diesel engine, the temperature of the gas at the end of the isobaric process.
There many not be a solution for some pairs of (T_2, T_3). That is, during some part of the cycle, the temperature, volume, pressure, or entropy may go out of the graphs' limits.

c. Analyze the results. If they are unusual, try deriving an expression(s) to explain the results.

3.4 **Efficiencies of Different Cycles**
Use the list of idealized engine cycles in Table 3.3 and rate the efficiencies of these cycles. Design reversible engines whose thermodynamic

Table 3.3: Idealized gas engine cycles

Engine cycle	Processes
Carnot cycle	Isothermal compression Adiabatic compression Isothermal expansion Adiabatic expansion back to The initial conditions
Brayton cycle	Isobaric compression Adiabatic compression Isobaric expansion Adiabatic expansion back to The initial conditions
Diesel cycle	Adiabatic compression Isobaric expansion Adiabatic expansion Isochoric process back to The initial conditions
Lenoir cycle	Isobaric compression Isochoric process (inc temp) Adiabatic expansion back to The initial conditions
Otto cycle	Adiabatic compression Isochoric process (inc temp) Adiabatic expansion Isochoric process (dec temp) Back to the initial conditions
Stirling cycle	Isothermal compression Isochoric process (inc temp) Isothermal expansion Isochoric process (dec temp) Back to the initial conditions

parameters stay within the limits of the plots. Whenever possible use the same upper and lower temperatures for each engine.
The Diesel, the Otto, and the Stirling cycles can easily be closed (i.e., returned to their initial conditions). Closing the other cycles requires some computation. Section 3.2.4 contains a discussion of the requirements for the Carnot engine.

3.5 **Irreversible Engines**
Select an irreversible engine and study the effects of heat loss on the engine's efficiency by varying the percentage of heat loss. Plot the efficiencies versus heat loss percentages for identical cycles.

3.6 **Design Your Own Engine Cycle**
Try to achieve an efficiency that is equal to or greater than the equivalent Carnot cycle, i.e., a Carnot cycle operating between the same upper and lower temperatures.

3.7 **Effects of Different Gases**
Choose an engine cycle and study the effects of changing the gas on the engine's efficiency. What accounts for the differences?

3.8 **Refrigerators**
Reverse the cycles in Table 3.3 and study the coefficient of performance of each cycle.

3.9 **More Realistic Irreversible Engines***
Try using more realistic models of irreversibility. (For examples, see refs. 3 and 4.)

3.5 Details of the Programs

ENGDRV is the driver program that can be used to call the programs: DIESEL, OTTO, WANKEL, and ENGINE (Design Your Own Engine). These programs also can be called directly by typing their names.

3.5.1 Input Screen Options for DIESEL, OTTO, and WANKEL

Maximum operating temperature of the engine	Integer between 1000 and 2000
Compression ratio	Number between 15 and 25 for Diesel engine Number between 5 and 12 for the Otto or Wankel engines

The compression ratio equals V_1/V_2, where V_1 is the initial volume and V_2 is the final volume after the adiabatic expansion. This data combined with the initial conditions (T = 300 K and P = 1 atm) is sufficient to define the engine's cycle.

3.5.2 Input Screen Options for ENGINE

Select engine type	Reversible engine Irreversible engine
Select gas	Helium (He) Argon (Ar) Nitrogen (N) Steam
Initial conditions	
Temperature (K)	A number between 300 and 2000 for He, Ar or N A number between 500 and 2000 for steam
Pressure (atm)	A number between 1 and 75
% of heat loss	A number between 0 and 99 (for irreversible engines only)
Output file name	The output disk file name The output file name only appears at the beginning of the program. Its default is ENGINE.DAT.

The following information maybe helpful in selecting variables. Helium and argon are monatomic molecules, nitrogen is a diatomic molecule, and steam is a triatomic molecule. Nitrogen's characteristics (specific heats, atomic mass) closely resemble those of air. The lower limit of the temperature is 300 K for helium, argon, and nitrogen, and 500 K for steam. The percentage of heat loss (irreversible engine only) is a number between 0 and 99 and is the percentage of heat lost during an isobaric, isochoric, or isothermal process.

During the operation of the program, there are radio buttons to select the process type (adiabatic, isobaric, isochoric, and isothermal) and sliders to change a thermodynamic variable (temperature, pressure, or volume). To the right of each slider, there is a box containing the current value of the variable. By clicking the mouse in the box, the user can type in a value for that variable. The user must press the **Enter** key to continue. The program checks that the inputted value of the thermodynamic variable does not cause any of the other thermodynamic variables to exceed their graphical limits.

The volume's upper limit is determined by the initial temperature.

3.5.3 Output Data File

The output data file includes the following:

- The engine type (reversible or irreversible).
- The gas (helium, argon, nitrogen or steam).
- The specific heat of the gas (J/K).
- The value of $\gamma = C_P/C_V$.
- The mass of a molecule of the gas (amu).
- A table listing the processes in the cycle, the gas's volume, pressure, temperature, entropy (J/K), the work done (J), the heat, and the change in the gas's internal energy going from the gas's previous state to its current state.
- Whether the cycle is an engine or a refrigerator.
- If the cycle is an engine, the total work done to the surroundings, the total heat absorbed by the gas, the heat absorbed by the gas, and the engine's efficiency.
- If the cycle is a refrigerator, the work done to the gas, the heat extracted from the colder heat source, and the coefficient of performance.

References

1. Brown, W. D. *Thermodynamics and Heat Engines.* London: Sir Isaac Pitman & Sons Ltd., 1964.

2. Cole, D. E. The Wankel Engine. *Scientific American* **227**:14–23, 1972.

3. De Vos, A. Reflections on the power delivered by endoreversible engines. *Journal of Phys D: Applied Physics* **20**:232–236, 1987.

4. Rubin, M. H. Optimal configuration of a class of irreversible heat engines. I and II. *Physical Review* A**19**:1272–1277, 1979.

5. Zemansky, M. W., Dittman, R. *Heat and Thermodynamics.* New York: McGraw-Hill, 1981.

4

Introduction to Probability and Statistics

L. B. Spornick

4.1 Introduction

The probability and statistics programs provide an introduction to probability theory and statistics which are part of the foundation for statistical mechanics and the kinetic theory of gases. They also are used to describe random walks, polymers, and the decay rate of radioactive atoms. (See refs. 1–4.) The central limit theorem, one of the most important theorems of probability theory, accounts for the fact that many phenomena in nature follow a Gaussian distribution. In this chapter we will discuss—

- Examples of some commonly used probability distributions: Gaussian (Normal), binomial (Bernoulli), Poisson, exponential, and Rayleigh distributions.
- An example of using the laws of probability to compute non-standard distributions (the Galton board).
- Application of the central limit theorem to the Galton board.
- An example of statistical independence of motion in two dimensions (random walks in two dimensions).
- An example of a mechanical system and its statistical model (the Kac ring).
- An example of a chaotic system (the stadium model).

4.2 Probability and Statistics

4.2.1 Probability

Probability theory originated in the 1650s from discussions between Blaise Pascal and Pierre Fernat about games of chance. In the early nineteenth century, Pierre Laplace and Karl Frederick Gauss realized that probability could be applied to many other areas. Probability theory can be as easily applied to the outcome of coin tosses as to the prediction of the number of radioactive decays. To compute the probability of an event, $P(\text{event})$, all the possible outcomes are enumerated. An event is a particular outcome, for example, obtaining a head from the toss of a coin. Assuming each possible outcome is equally likely, $P(\text{event})$ is defined as the number of times that a particular event occurs in the enumeration, $n(\text{event})$, divided by the total number of possible outcomes n:

$$P(\text{event}) = \frac{n(\text{event})}{n}. \tag{4.1}$$

$P(\text{event})$ has the following properties:

$$0 \leq P(\text{event}) \leq 1 \tag{4.2}$$

and

$$\sum_{\text{all events}} P(\text{event}) = 1. \tag{4.3}$$

Statisticians define a *random variable* as an event that occurs with a known probability. There are two types of random variables: *discrete random variables*, such as the number of radioactive atoms that have decayed in a given time period, and *continuous random variables*, such as molecular velocities.

The *probability distribution function*, $P(\text{event})$, provides a convenient means of representing the probabilities associated with discrete random variables. The *cumulative distribution function*, $F(x)$, and its derivative, the *probability density function*, $f(x)$, provide a convenient means for representing the probability associated with continuous random variables. $F(x)$ is the probability that the random variable has a value between $-\infty$ and x :

$$F(x) = \int_{-\infty}^{x} f(x)\, dx. \tag{4.4}$$

The area under the $f(x)$ curve between any two values x_1 and x_2 equals the probability that the random variable will lie between these values:

$$P(x_1 \leq x \leq x_2) = \int_{x_1}^{x_2} f(x)\, dx. \tag{4.5}$$

4.2.2 The Laws of Probability

There are two primary laws of probability theory: the *law of addition* and the *law of multiplication*. These two laws can be used to compute the probability distribution function for a sequence of events.

The law of addition states that if A and B are mutually exclusive events, the probability that either A or B occurs ($P(A \text{ or } B)$) is equal to the sum of the probability that A occurs, $P(A)$, and the probability that B occurs, $P(B)$:

$$P(A \text{ or } B) = P(A) + P(B). \tag{4.6}$$

Events A and B are mutually exclusive if the fact that one occurs precludes the occurrence of the other. An example of two mutually exclusive events is throwing a 1 or 2 with a die:

$$P(1 \text{ or } 2) = P(1) + P(2). \tag{4.7}$$

If A and B are not mutually exclusive, the law of addition becomes

$$P(A \text{ or } B) = P(A) + P(B) - P(A \text{ and } B), \tag{4.8}$$

where $P(A \text{ and } B)$ is the probability that both A and B occur simultaneously. An example of events that are not mutually exclusive is $A = \{2, 4, 6\}$, the even faces of a die, and $B = \{1, 2, 3\}$, the faces of a die that are less than four. Both A and B include 2 (see Fig. 4.1).

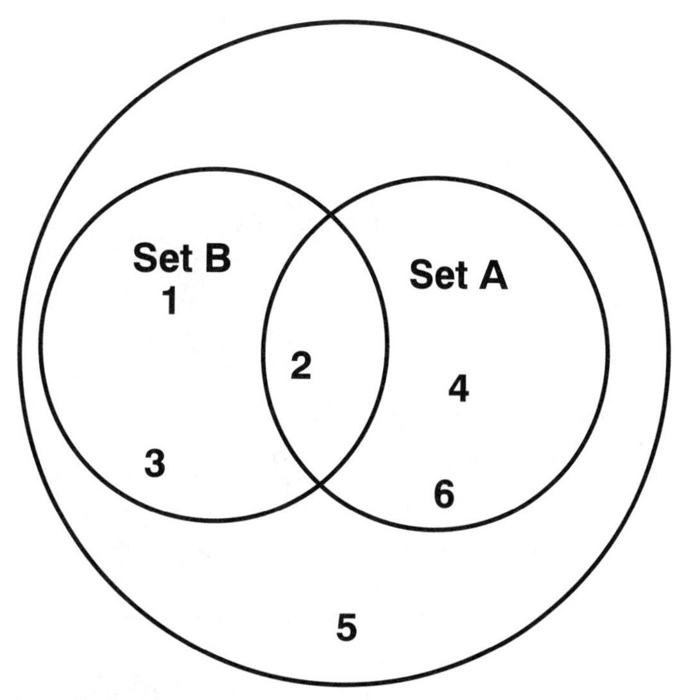

Figure 4.1: The set of outcomes of a die's roll.

$$P(A \text{ or } B) = P(2) + P(4) + P(6) + P(1) + P(2) + P(3) - P(2) \qquad (4.9)$$

where $P(A) = P(2) + P(4) + P(6)$, $P(B) = P(1) + P(2) + P(3)$, and $P(A$ and $B) = P(2)$.

The law of multiplication states that the probability that both events A and B will occur, $P(A$ and $B)$, is equal to the probability that A occurs times the probability that B occurs, if A and B are statistically independent events

$$P(A \text{ and } B) = P(A)P(B). \qquad (4.10)$$

Events are statistically independent if the probability that one event occurs is independent of whether the other has occurred. The outcomes of two tosses of a fair die is an example of two statistically independent events. The probability that the second toss produces a particular outcome is independent of the outcome of the first toss.

If the events are statistically dependent, the law of multiplication becomes

$$P(A \text{ and } B) = P(A)P(B|A) = P(B)P(A|B) \qquad (4.11)$$

where $P(B|A)$ is the conditional probability that B will occur given that A has occurred

$$P(B|A) = n(\text{events common to } A \text{ and } B)/n(A). \qquad (4.12)$$

Using the above example, because '2' is the only element common to both A and B, $P(B/A) = 1/3$, $P(A) = 1/2$, and

$$P(A \text{ and } B) = 1/6 \qquad (4.13)$$

4.2.3 Means and Moments

The mean (or average) of a function $g(x)$ is defined as

$$\langle g(x) \rangle = \sum_x g(x)P(x) \qquad (4.14)$$

if the events are discrete, or

$$\langle g(x) \rangle = \int_{-\infty}^{\infty} g(x)f(x)\,dx \qquad (4.15)$$

if the events are continuous. $P(x)$ is the probability distribution function and $f(x)$ is the probability density function.

The nth moment about the mean of x is

$$\langle (x - \langle x \rangle)^n \rangle = \sum_x (x - \langle x \rangle)^n P(x) \qquad (4.16)$$

if the events are discrete, or

$$\langle (x - \langle x \rangle)^n \rangle = \int_{-\infty}^{\infty} (x - \langle x \rangle)^n f(x)\, dx \tag{4.17}$$

if the events are continuous.

The first moment about the mean of x is always zero:

$$\begin{aligned} \langle x - \langle x \rangle \rangle &= \sum_x (x - \langle x \rangle) P(x) \\ &= \sum_x x P(x) - \langle x \rangle \sum_x P(x) \\ &= \langle x \rangle - \langle x \rangle = 0. \end{aligned} \tag{4.18}$$

The second moment about the mean of x is call the *variance*, $\langle (x - \langle x \rangle)^2 \rangle$, and the square root of the variance is called the *standard deviation*, $\sqrt{\langle (x - \langle x \rangle)^2 \rangle}$.

4.2.4 Common Probability Distributions

The Binomial Distribution. The binomial (or Bernoulli) distribution was developed by James Bernoulli around 1700. It provides the probability of getting k successes in N statistically independent trials and describes processes such as random walks and coin tosses.

To illustrate the binomial distribution, consider a sequence of four tosses of a coin. Let us define P_h as the probability that a head appears and $P_t = (1 - P_h)$ as the probability that a tail appears. If we do not care about the order in which the heads and tails appear, the probability of obtaining two heads and two tails is

$$P(2 \text{ heads, } 2 \text{ tails}) = \frac{4!}{2!\,2!} P_h^2 P_t^2, \tag{4.19}$$

where $4!/(2!\,2!) = 6$ is the number of ways of obtaining two heads and two tails in four tosses (tthh, thth, thht, hhtt, htht, htth, where h represents a "head" and t represents a "tail"). We have summed all the probabilities associated with the mutually exclusive outcomes in which two heads and two tails appear. The probability of obtaining any one of these is $P_h^2 P_t^2$.

In general, the probability of n heads and $(N - n)$ tails in N tosses is

$$P(N, n) = C(N, n)\, P_h^n\, P_t^{(N-n)}, \tag{4.20}$$

where

$$C(N, n) = \frac{N!}{n!\,(N - n)!}. \tag{4.21}$$

$P(N, n)$ is the binomial distribution associated with obtaining n heads in N tosses (trials). $C(N, n)$ equals the number of unique ways of arranging n heads and

$(N - n)$ tails and is known as the binomial coefficient because it also appears in the expression for the expansion of a binomial raised to the power N:

$$(a + b)^N = C(N, N)a^N b^0 + C(N, N - 1)a^{N-1}b^1 + \cdots + C(N, 0)a^0 b^N . \tag{4.22}$$

If we let $a = P_h$ and $b = P_t = (1 - P_h)$ and use Eq. 4.22, we can show that the binomial distribution is normalized to unity:

$$\sum_n P(N, n) = (P_h + (1 - P_h))^N = 1 . \tag{4.23}$$

The mean of n is

$$\langle n \rangle = \sum_n n\, P(N, n) = NP_h \tag{4.24}$$

and the variance is

$$\langle (n - \langle n \rangle)^2 \rangle = \sum_n (n - \langle n \rangle)^2 P(N, n) = NP_h(1 - P_h) . \tag{4.25}$$

If we change head to right and tail to left, the binomial distribution describes the outcome of a one-dimensional random walk of N equal-length steps. Variations on the random walk provide models for the diffusion of a molecule in a gas, Brownian motion, the characterization of long polymer chains, and the estimation of definite integrals.

The Gaussian Distribution. The Gaussian distribution is probably the most important distribution in statistics. It was discovered in 1733 by Abraham de Moivre as an approximation to the binomial distribution in the limit that the number of trials goes to infinity. In 1809 Carl Frederick Gauss derived it as the law of errors of observations. He showed empirically that errors in astronomical observation follow the Gaussian curve with great accuracy. The real importance of the Gaussian distribution comes from its connection to the central limit theorem.

The probability density function associated with the Gaussian distribution is

$$f(x) = \frac{1}{\sigma\sqrt{2\pi}} e^{-(x-\mu)^2/2\sigma^2} , \tag{4.26}$$

where $-\infty < x < \infty$ is a continuous random variable, μ is the mean, and σ is the standard deviation. The probability that the random variable x has a value between x_1 and x_2 is

$$P(x_1 < x < x_2) = \frac{1}{\sigma\sqrt{2\pi}} \int_{x_1}^{x_2} e^{-(x-\mu)^2/2\sigma^2}\, dx . \tag{4.27}$$

The Central Limit Theorem and the Chi-Square Test. The central limit theorem states that the distribution of a random variable which is the sum of a large number of statistically independent random variables approaches the Gaussian distribution. This theorem applies regardless of the distributions that describe the statistically independent random variables. For example, the computed average height of a class of eighth grade students is a random variable and the distribution of these random variables for the eighth-grade classes in a state approaches the Gaussian distribution.

By comparing the distribution of measurements to a Gaussian distribution, the chi-square criterion provides a test of the central limit theorem. This test involves the following steps:

1. Take n measurements.

2. Group the measurements into k bins of width δ.

3. Compute the chi-square distribution of $(k - h - 1)$ degrees of freedom (h is the number of fitted parameters). If the mean and variance are known, h is 0. If the mean and variance are unknown, $h = 2$ and the mean and variance are approximated by the measurements' mean and variance.

4. Determine the goodness of fit from the calculated chi-square distribution and the tabulated chi-square probabilities.

The chi-square distribution of $(k - h - 1)$ degrees of freedom is given by

$$\chi^2 = \sum_i \frac{(n_i - E_i)^2}{E_i^2}, \tag{4.28}$$

where n_i is the number of measurements in the ith bin, and E_i is the expected number of measurements in the ith bin:

$$E_i = n \int_{x_i-\delta/2}^{x_i+\delta/2} f(x)\, dx, \tag{4.29}$$

where

$$n = \sum n_i . \tag{4.30}$$

Selecting the best bin size is important. In testing the central limit theorem, $f(x)$ is the Gaussian probability density function (Eq. 4.26).

The relevant quantity in determining the goodness of fit is the integral of the chi-square probability density function, $P(\chi^2, N)$, where N is the number of degrees of freedom. $P(\chi^2, N)$ is the probability, given that $f(x)$ describes the data, of obtaining a χ^2 as large, or larger, than the one calculated. $P(\chi^2, N)$ has been tabulated and can be found in many probability references. (See Table 4.1 for values with $N = 12$ and 27.) If χ^2 is significantly larger than $P(\chi^2, N)$, it indicates that $f(x)$ is probably not a good description of the data. If $\chi^2 = 0$, it indicates that

Table 4.1: Chi-square table

Degree of freedom N	$P = 0.99$	0.98	0.95	0.90	0.80	0.70	0.50	0.30	0.20	0.10	0.05	0.01
12	3.6	4.2	5.2	6.3	7.8	9	11	14	16	19	21	26
27	13	14	16	18	21	23	26	30	33	37	40	47

$f(x)$ describes the data perfectly. For example, if there are 12 degrees of freedom and $\chi^2 = 6.2$, it means that there is a 90 percent probability that $f(x)$ gives the correct description of the data.

The Rayleigh Distribution. The Rayleigh distribution was originally derived in 1919 by Lord Rayleigh while working on a problem in acoustics. The Rayleigh distribution describes the radial distribution of two or more Gaussian distributed, statistically independent random variables that have zero mean and the same variance. The joint cumulative density function for two statistically independent, Gaussian distributed random variables with zero mean and the same variance is

$$f(x,y) = \frac{1}{2\pi\sigma^2} e^{-(x^2+y^2)/2\sigma^2} . \tag{4.31}$$

The radial distribution is computed from the joint probability by converting the variables from (x,y) to (r,θ) and integrating over θ:

$$\begin{aligned} f(x,y)\,dx\,dy &= f(r,\theta)\,rdr\,d\theta \\ &= \frac{1}{2\pi\sigma^2} e^{-r^2/2\sigma^2}\, r\,dr\,d\theta \end{aligned} \tag{4.32}$$

$$f(r)\,dr = \int_0^{2\pi} f(r,\theta)\, r\,dr\,d\theta \tag{4.33}$$

$$\begin{aligned} &= \frac{1}{2\pi\sigma^2} e^{-r^2/2\sigma^2}\, r\,dr \int_0^{2\pi} d\theta \\ &= \frac{1}{\sigma^2} r e^{-r^2/2\sigma^2}\, dr . \end{aligned} \tag{4.34}$$

The Rayleigh distribution's probability density function in n dimensions is

$$f(r,n) = 2r^{n-1} \frac{1}{2\sigma^2} e^{-r^2/2\sigma^2} \Gamma(\frac{n}{2}), \tag{4.35}$$

where $r = \sqrt{x_1^2 + x_2^2 + \cdots + x_n^2}$ and

$$\Gamma(n) = \int_0^{\infty} e^{-x}\, x_{n-1}\, dx . \tag{4.36}$$

Tabulated values of $\Gamma(n)$ can be found in mathematical handbooks.

The Poisson and Exponential Distributions. The Poisson distribution was discovered in 1837 by S. D. Poisson. It is the limiting form of the binomial distribution when there is a large number of trials but a small probability of success. The exponential distribution can be regarded as the continuous analog of the Poisson distribution.

The classic example of a system described by the Poisson distribution is the number of errors on a printed page. The probability of finding one error is small, assuming that the proofreader did his/her job correctly. The probability of finding two or more errors on a page is almost nonexistent. The Poisson distribution also describes the distribution of the number of decays in a unit time for an ensemble consisting of a large number of similar samples of radioactive atoms. The exponential distribution is useful in studying decay times of radioactive particles, in particular, the half-life of radioactive elements. The exponential distribution also can be used to estimate the number of radioactive atoms as a function of time.

The Poisson distribution is given by

$$P(n) = \frac{\mu^n}{n!}e^{-\mu}, \tag{4.37}$$

where n is the number of successes, and μ is both the mean number of successes and the square of the standard deviation. The Poisson distribution is the limiting form of the binomial distribution $P(N, n)$:

$$\lim_{P_s \to 0, N \to \infty} P(N, n) = \lim_{P_s \to 0, N \to \infty} P_s^n (1 - P_s)^{N-n} \frac{N!}{n!(N-n)!} \tag{4.38}$$

$$= \lim_{P_s \to 0, N \to \infty} \frac{(NP_s)^n}{n!}\left(1 - \frac{NP_s}{N}\right)^{N-n} \left(\frac{N}{N}\right) \cdots \left(\frac{N-n+1}{N}\right)$$

$$= \frac{\mu^n}{n!}e^{-\mu} = P(n), \tag{4.39}$$

where N is the number of trials, n is the number of successes, P_s is the probability of a success, $\mu = NP_s$, and

$$\lim_{N \to \infty} \frac{N}{N} \cdots \frac{N-n+1}{N} \longrightarrow 1$$

$$\text{and} \lim_{N \to \infty} (1 - P_s)^{N-n} \longrightarrow e^{-NP_s} = e^{-\mu}.$$

The probability density function associated with the exponential distribution is given by

$$f(x) = \lambda e^{-\lambda x}, \tag{4.40}$$

where $x \geq 0$ and the mean and the standard deviation equals $1/\lambda$. The cumulative probability function is

$$P(x_1 \leq x) = \int_0^{x_1} \lambda e^{-\lambda x'} dx' = 1 - e^{-\lambda x_1}. \tag{4.41}$$

The exponential distribution describes the radioactive decay law. The process of radioactive decay is a one-shot process, meaning that a radioactive nucleus can decay only once. As long as a particular nucleus has not decayed, the probability that it will decay in the next unit of time is a constant. Therefore, the probability that a nucleus will decay in the time interval between t and $t + dt$ is proportional to dt. For a sample of N radioactive nuclei, the number of atoms that will decay in a unit time is proportional to N:

$$-dN = \lambda N dt, \tag{4.42}$$

where λ is called the decay constant. The minus sign in Eq. 4.42 indicates that the number of radioactive nucleus decreases with time. The solution to Eq. 4.42 is

$$N(t) = N_0 e^{-\lambda t}, \tag{4.43}$$

where $N(t)$ is the number of radioactive nuclei remaining at time t and N_0 is the initial number of radioactive nuclei.

The half-life of a radioactive nuclei is defined as the time during which the number of radioactive nuclei reduces to one-half the original number:

$$\frac{1}{2}N_0 = N_0 e^{-\lambda t_{1/2}} \tag{4.44}$$

or

$$t_{1/2} = \frac{1}{\lambda} \ln 2 . \tag{4.45}$$

4.2.5 Generalized Probability Distributions

Many probability distributions cannot be written in a compact form such as the Gaussian distribution or the binomial distribution. For discrete random variables, a generalized probability distribution can be represented by a table in which the random variable(s) is in one (or more) column(s) and the corresponding probability is in another. The probability of finding a particular value of a random variable equals the sum of the probabilities associated with all of the unique paths to that random variable. The Galton board provides a simple illustration of how to determine the probability distribution.

The Galton Board. A Galton board is a long, narrow board with an opening at the center of the top, a triangular grid of pins, traps and/or reflecting/absorbing walls just below the opening and a series of bins at the bottom. Balls are dropped into the board from the opening and are deflected by the pins and/or reflecting walls or are absorbed by the traps or absorbing walls. The final distribution of the balls in the bins is a function of the deflection probability (the probability that the ball will go to the right) of each pin and the positions of the traps, reflecting walls, and absorbing walls. (See ref 2.)

Consider a Galton board with two levels and no traps or reflecting walls or absorbing walls (see Fig. 4.2). There is only one path to bin 0 (the leftmost bin):

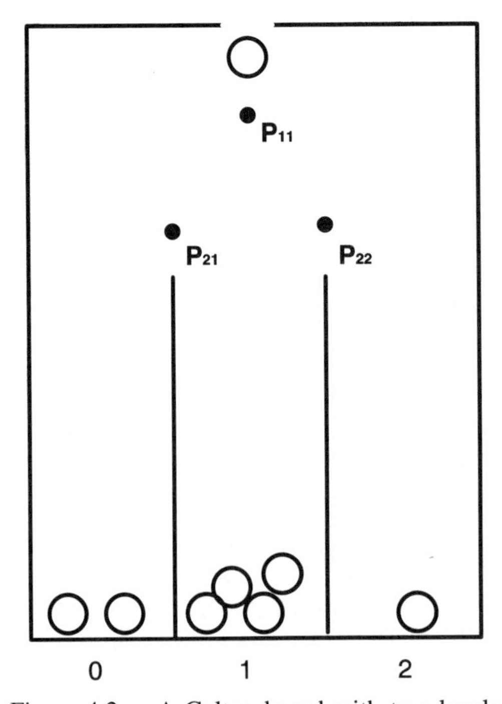

Figure 4.2: A Galton board with two levels.

pin(1,1) $\longrightarrow$ pin(2,1) $\longrightarrow$ bin 0. Using the law of multiplication, the probability of a ball falling into bin 0 is

$$(1 - P_{11})(1 - P_{21}) \tag{4.46}$$

P_{ij} is the probability that the ball hitting pin (i, j) will go to the right and $(1 - P_{ij})$ is the probability that the ball will go to the left. Pin (i, j) is the jth pin from the left on the ith row. If pin (i, j) is a trap or part of an absorbing wall, P_{ij} and $(1 - P_{ij}) = 0$. If pin (i, j) is part of a reflecting wall, $P_{ij} = 0$ if the wall is on the right side of the board or $P_{ij} = 1$ if it is on the left side.

There are two paths to bin 1: pin(1,1) $\longrightarrow$ pin(2,1) $\longrightarrow$ bin 1 and pin(1,1) $\longrightarrow$ pin(2,2) $\longrightarrow$ bin 1. The probability associated with the first path is

$$(1 - P_{11})P_{21} \tag{4.47}$$

and the probability associated with the second path is

$$P_{11}(1 - P_{22}) \,. \tag{4.48}$$

So using the law of addition, the total probability associated with bin 1 is

$$(1 - P_{11})P_{21} + P_{11}(1 - P_{22}) \,. \tag{4.49}$$

There is only one path to bin 2: pin(1,1) $\longrightarrow$ pin(2,2) $\longrightarrow$ bin 2. The probability associated with this path is $P_{11}P_{22}$. The probability distribution for this Galton board is given in Table 4.2.

Table 4.2: Probability distribution for a two-level Galton board

Random variable	Probability
Bin 0	$(1 - P_{11})(1 - P_{21})$
Bin 1	$(1 - P_{11})P_{21} + P_{11}(1 - P_{22})$
Bin 2	$P_{11}P_{22}$

Clearly, this exercise is a simple example, but more complicated probability distributions are calculated in exactly the same way.

4.2.6 Kinetic Theory of Gases and Statistical Mechanics

Statistics is the process of gathering, analyzing, and making inferences from numerical data. The kinetic theory of gases and statistical mechanics combine statistical methods and the laws of classical mechanics and/or quantum mechanics to describe the macroscopic properties of matter in terms of the most probable behavior of the system. One reason for the need for statistical methods to describe mechanical systems is that it is not possible to solve the equations of motion for many body systems. Another reason is that there exist chaotic systems whose equations of motion can be solved numerically, but, because of extreme sensitivity to initial conditions, must be described by statistical methods.

When Boltzmann first introduced the kinetic theory of gases, a number of objections were raised. In particular, systems described by the Boltzmann H theorem do not exhibit time reversal or have a Poincaré cycle. Poincaré had shown that after a sufficiently long time, every finite mechanical system returns arbitrarily close to its initial state. This time period is known as the Poincaré cycle. Since for realistic systems, the Poincaré cycle is much larger than the lifetime of the universe, it is important only for fundamental reasons, not realistic systems. However, in the mid-nineteenth century, when Boltzmann first introduced his H theorem, it was a major concern.

The Kac ring was developed by Marc Kac to model a finite mechanical system whose equations of motion are solvable, has time reversibility, has a Poincaré cycle, and can be described by a statistical model. It provides a comparison between a physical system and its statistical description.

The Kac Ring. A Kac ring consists of a ring with a number, n, of evenly spaced points and balls around the perimeter (see Fig. 4.3). At m of the points are markers. The balls move around the ring and are either dark or light. The equation of motion is, that as the balls move around the ring, they change color when they pass a point with a marker. It is a deterministic model that demonstrates time reversal and has a Poincaré cycle. Yet, for large rings ($n >> 1$), there is good agreement between the Kac ring and the corresponding canonical ensemble, i.e., the statistical description of the Kac ring.

The canonical ensemble of the Kac ring is an ensemble of rings. Each ring has the same number of points and markers and a unique distribution of markers. For a Kac ring, the quantity of interest is the number of dark balls $N(D)$ minus the number of light balls $N(L)$. The system is in equilibrium when $N(D) - N(L) = 0, 1,$ or -1. Why? There are two reasons: 1) for a large ring, after many turns, the probability of a ball being either light or dark is equal; and 2) there are more unique arrangements satisfying the requirement that $(N(D) - N(L)) = 0, 1,$ or -1 than any other condition (see Eq. 4.21). Therefore, systems for which $N(D) - N(L) = 0, 1,$ or -1 occupy the largest volume in phase space and $N(D) - N(L) = 0, 1,$ or -1 defines equilibrium. For each turn of the ring (i.e., when a ball passes an adjacent point), $N(D) - N(L)$ is calculated exactly and compared to the canonical ensemble average $(\langle N(D) - N(L) \rangle)$.

Statistical Methods and Chaos. Classical mechanics assumes that if a system's equations of motion are solvable, its properties are predictable. These systems are called well-defined or stable. There exists, however, a large number of systems whose equations of motion are solvable, but, because of extreme sensitivity to ini-

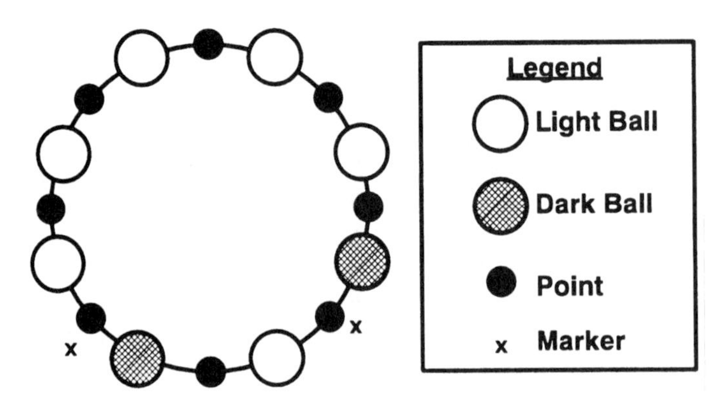

Figure 4.3: An example of a Kac ring.

tial conditions, have properties that are not predictable over long periods of time. (See refs 7 and 8.) These systems are called chaotic. Chaos only occurs in nonlinear systems. Chaotic systems can be described using statistical methods. The following discussion provides an introduction to the nomenclature of chaos.

There are two types of chaotic systems: dissipative dynamical systems and conservative systems. In a dissipative dynamical system, the system's energy varies due to friction and driving forces. In a conservative system, the system's energy is constant. Dissipative dynamical systems can have attractors and conservative systems can not. An attractor is a bounded region in phase space to which the phase trajectory becomes attracted in the course of time.

Phase space is a $6N$-dimensional space where N is the number of particles in the system. The axes are the $3N$ configuration space coordinates and their corresponding $3N$ conjugate coordinates. Usually, the configuration space coordinates consist of the $3N$ Cartesian coordinates and the corresponding conjugate coordinates consist of the $3N$ momentum coordinates.

A phase trajectory is the path that a system's phase point travels in time. A system's phase point at a particular time is specified by the system's state, i.e., the values of the $6N$ coordinates. For example, the phase point at time t for a system consisting of two gas atoms in a box is

$$(x_1(t), y_1(t), z_1(t), x_2(t), y_2(t), z_2(t), p_{x_1}(t), p_{y_1}(t), p_{z_1}(t), p_{x_2}(t), p_{y_2}(t), p_{z_2}(t)),$$

where $x_1(t)$ is the x coordinate of the first particle at time $t, \ldots, p_{x_1}(t)$ is the x axis momentum of the first particle at time $t, \ldots$.

The phase trajectory contains the complete history of the system. Because phase space usually has six or more dimensions, it is difficult to visualize the phase trajectory. Instead, it is convenient to observe where the phase trajectory intersects a plane in phase space. If this plane is properly chosen, the plot of points (called a Poincaré section) has the same properties as the phase trajectory. For example, if there is an attractor in phase space, there also is one on the plane.

Both dissipative dynamical chaotic systems and conservative chaotic systems have a complicated time evolution that is characterized by a Lyapunov exponent and a "predictability horizon." The Lyapunov exponent measures how fast nearest neighbor trajectories exponentially diverge. The predictability horizon is defined as the time beyond which nearest-neighbor trajectories diverge randomly. Nearest neighbor trajectories are trajectories whose initial conditions differ by one unit in the last (right-most) decimal place.

For non-chaotic systems, not only are the equations of motion solvable, but nearest-neighbor trajectories maintain a well-defined relationship between them so there is no predictability horizon. An example of a well-defined system is a particle of mass m moving in a uniform gravitational field g. The equations of motion are

$$\begin{aligned} x(t) &= x_0 + v_{x0}t \\ p_x(t) &= mv_{x0} \\ y(t) &= y_0 + v_{y0}t - \frac{1}{2}gt^2 \\ p_y(t) &= m(v_{y0} - gt), \end{aligned} \tag{4.50}$$

where (x_0, y_0) and (v_{x0}, v_{y0}) are the initial position and velocity, respectively. The relationship between the nearest-neighbor trajectories is

$$\begin{aligned}
\Delta x(t) &= \Delta x_0 + \Delta v_{x0} t, \\
\Delta p_x(t) &= m \Delta v_{x0} \\
\Delta y(t) &= \Delta y_0 + (\Delta v_{y0}) t \\
\Delta p_y(t) &= m \Delta v_{y0}
\end{aligned} \tag{4.51}$$

where Δx_0, Δy_0, Δv_{x0}, and Δv_{y0} are the differences in the initial conditions between the nearest neighbors.

For a chaotic system, after the predictability horizon, the distances between nearest-neighbor trajectories in phase space as a function of time becomes random. Decreasing the initial distance between nearest neighbor trajectories by increasing the accuracy of the measurements or the precision of the computer's arithmetic only changes the predictability horizon slightly because the divergence is exponential.

The Stadium Model. The stadium model illustrates a chaotic system whose equation of motion is solvable. The model consists of a ball confined to move in a stadium-shaped container (a rectangle capped by two semicircles of unit radius). The ball makes elastic collisions with the stadium's wall. It is a good representation of a two-dimensional ideal gas atom. The system's equation of motion is simply that the angle of incidence equals the angle of reflection. The ball's speed remains constant. The stadium model is a conservative system. (See ref. 9)

Since the ball's speed is constant, its trajectory can be specified by giving the ball's position and direction of motion when it hits the stadium wall. The position is specified by its location on the stadium's border, s, and the direction of motion, p, is specified by the cosine of the angle of incidence, α (see Fig. 4.4). The quantity s can vary from 0 to $2(\pi +$ the length of the straight part of the stadium). (s, p) are conjugate coordinates. The ball's path can be described by a series of numbered pairs of $(s(n), p(n))$, where n is the number of bounces.

The ball's trajectory is either periodic or aperiodic (see Fig. 4.5). Periodic trajectories are well-defined and have closed (repeating) trajectories in phase space. Aperiodic trajectories are chaotic. The chaotic behavior manifests itself in our inability to predict the ball's trajectory beyond a predictability horizon. Most of the ball's trajectories are chaotic because the curved boundary reflects the ball in such a way that nearest-neighbor trajectories diverge randomly within a few bounces. Periodic and aperiodic trajectories may be infinitesimally close.

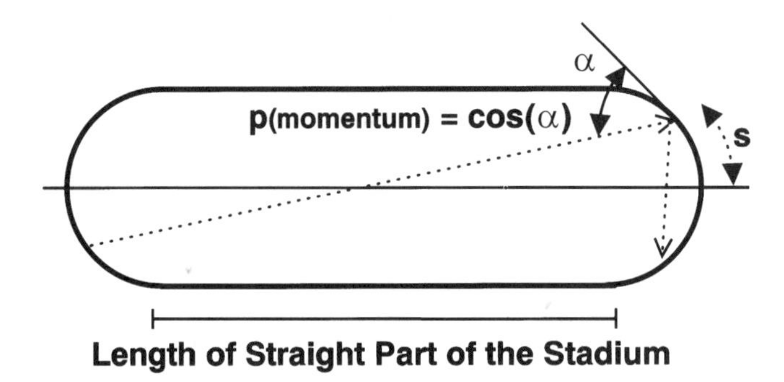

Figure 4.4: The stadium model.

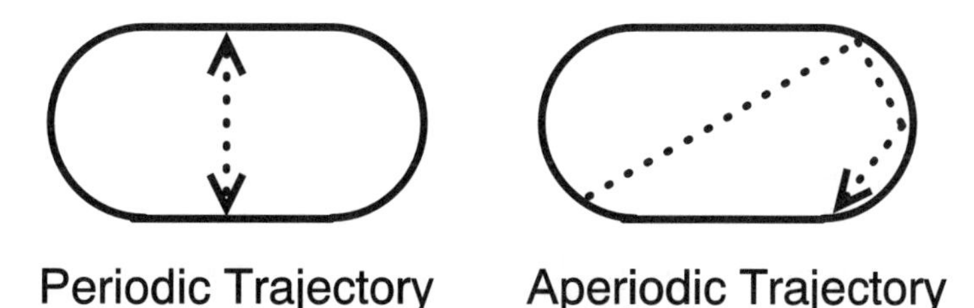

Figure 4.5: Periodic versus aperiodic trajectory.

4.3 Computational Approach

4.3.1 The Models

Introduction to probability and statistics contains five programs: GALTON, (Galton board), POISEXP (Poisson and exponential distributions), TWOD (random walks in two dimensions), KAC (Kac ring), and STADIUM (stadium model). The first three programs provide examples of some of the commonly used probability distributions (Gaussian, Binomial, Poisson, Exponential, Chi-square, and Rayleigh), an example of a generalized probability distribution, and a test of the Central limit theorem. These programs use a random number generator with a randomly selected seed to produce random events. KAC illustrates the difference between a Kac ring and its statistical model. STADIUM is an example of a system that has both well-defined and chaotic behavior.

4.3.2 The Galton Board Model

GALTON simulates a Galton board (see section 4.2.5). The program has three options:

- A traditional Galton board in which all the pins have the same deflection probability.

- A customized Galton board in which the user can place traps, reflecting walls, and absorbing walls (an absorbing wall is a column of traps), as well as designate the individual pins' deflection probabilities.

- A test of the central limit theorem.

The results of a traditional Galton board are compared to the corresponding binomial and Gaussian distributions. The results of a customized Galton board are compared to the corresponding generalized probability distribution that is described in section 4.2.5.

The test of the central limit theorem (see section 4.2.4) is an option that can be accessed after a Galton board has been defined. The test consists of running the defined Galton board 250 times and comparing the distribution of the computed 250 means to a Gaussian distribution whose mean and standard deviation is the same as the mean and standard deviation of the computed 250 means. According to the central limit theorem, the distribution of the computed 250 means should follow a Gaussian distribution (the $f(x)$ in eq. 4.29) with the same mean and standard deviation as the distribution of the computed 250 means. The chi-square criterion is used to demonstrate the central limit theorem.

4.3.3 The Poisson and Exponential Distribution Model

POISEXP simulates samples of 750 radioactive atoms with a known half-life ($t_{\frac{1}{2}}$). The method used to determine if an atom has decayed within a specified time interval (the observation time t_{obs})* involves generating a random number r and comparing it to the probability of a decay ($t_{obs} \times \ln(2)/t_{\frac{1}{2}}$ for the Poisson distribution or $t_{obs} \ln(2)/$(number of observations) for the exponential distribution). If r is greater than the probability of a decay, the atom decays.

There are two options:

- An illustration of the Poisson distribution.

- An illustration of the exponential distribution.

For the Poisson distribution option, the distribution of the number of decays for an ensemble of samples is plotted. This distribution is compared to the corresponding Poisson distribution. For the exponential distribution option, the number of radioactive atoms as a function of time for one sample is compared to the corresponding exponential distribution.

4.3.4 Model of Random Walks in Two Dimensions

TWOD simulates a two-dimensional unbiased random walk in which the "drunk" takes a specified number of steps. The drunk can move either on a grid or freely on the plane. The walk is repeated a specified number of times and the distributions of the final x direction displacement, the final y direction displacement and the final radial displacement are plotted. The final x and y displacements distributions are compared to a binomial distribution. The final radial displacement distribution for the random walk on a plane is compared to a Rayleigh distribution.

4.3.5 Kac Ring

KAC simulates a Kac ring (see section 4.2.6). There are two options:

- A small Kac ring (9 balls).

- A large Kac ring (2001 balls).

*For the Poisson distribution, the unit of time of t_{obs} is minutes and for the exponential distribution it is half-lives.

The small Kac ring makes two complete rotations (18 turns or one Poincaré cycle), then reverses its direction and makes two more complete rotations. The plots of $N(D) - N(L)$ and $\langle N(D) - N(L)\rangle$ illustrate the ring's time reversal and Poincaré cycle. The large Kac ring has two parts. In the first part, the ring makes one complete rotation (half a Poincaré cycle). $N(D) - N(L)$ is plotted as the ring turns. The average (mean) and the standard deviation σ of $N(D) - N(L)$ are calculated in this part. In the second part, the ring makes a smaller number of turns and $N(D) - N(L)$ and $\langle N(D) - N(L)\rangle$ are plotted as the ring turns. This plot demonstrates that as the ring becomes larger, the agreement between $N(D) - N(L)$ and $\langle N(D) - N(L)\rangle$ improves.†

4.3.6 The Stadium Model

STADIUM simulates the stadium model (see section 4.2.6). The program has two options:

- A one-ball model.
- A two-ball model.

The one-ball model investigates the question, "At what point does the computer's finite precision arithmetic with round-off errors affect the resulting numerical answer?" In the one-ball model, the ball makes n bounces, reverses its momentum, and makes n more bounces. If the numerical inaccuracies are insignificant, the ball retraces its path ending at the same place it began. If numerical inaccuracies are significant, the ball's final position differs from its initial position. When the difference between the initial position and the final position becomes significant, it is not possible to determine if the system is chaotic for those initial conditions and that number of bounces.

For the one-ball model, the computer screen is divided into two sections (see Fig. 4.6). The stadium's outline and the ball's trajectory appear on the left side of the screen. The first n bounces appear in white and the last n bounces appear in red. A plot of the position where the ball hits the outline versus the cosine of the angle of incidence, p, appears on the right side of the screen. This plot is a Poincaré section.

The two-ball model computes the predictability horizons for nearest-neighbor trajectories in which both initial conditions (the ball's starting location and the ball's trajectory's initial angle θ_i with respect to the x axis, see Fig. 4.7) are 0.1, 0.01, 0.001, and 0.0001 units apart.

For the two-ball model, the computer screen is divided into two sections (see Fig. 4.7). The stadium's outline and the balls' trajectories appear on the left side of the screen. The first ball's trajectory appears in white and the second ball's trajectory appears in red. Plots of the phase space separation of the nearest-neighbor trajectories, $d_n = ((s_1(n) - s_2(n)) + (p_1(n) - p_2(n))^2)^{1/2}$, as a function of n and the initial nearest-neighbor separation appear on the right side of the screen.

† Because the number of members in the canonical ensemble is very large, an approximation to the canonical ensemble average is used. This approximation assumes that the number of turns is much less than the number of markers, which in turn is much less than the number of balls. (See References 5 and 6.)

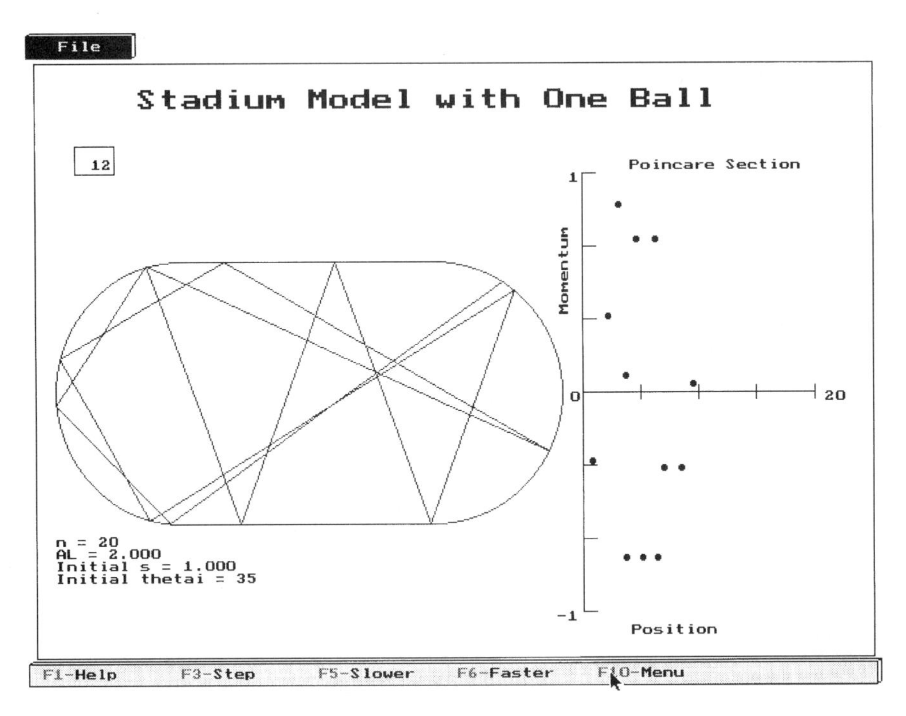

Figure 4.6: The screen for the one-ball model after twelve bounces.

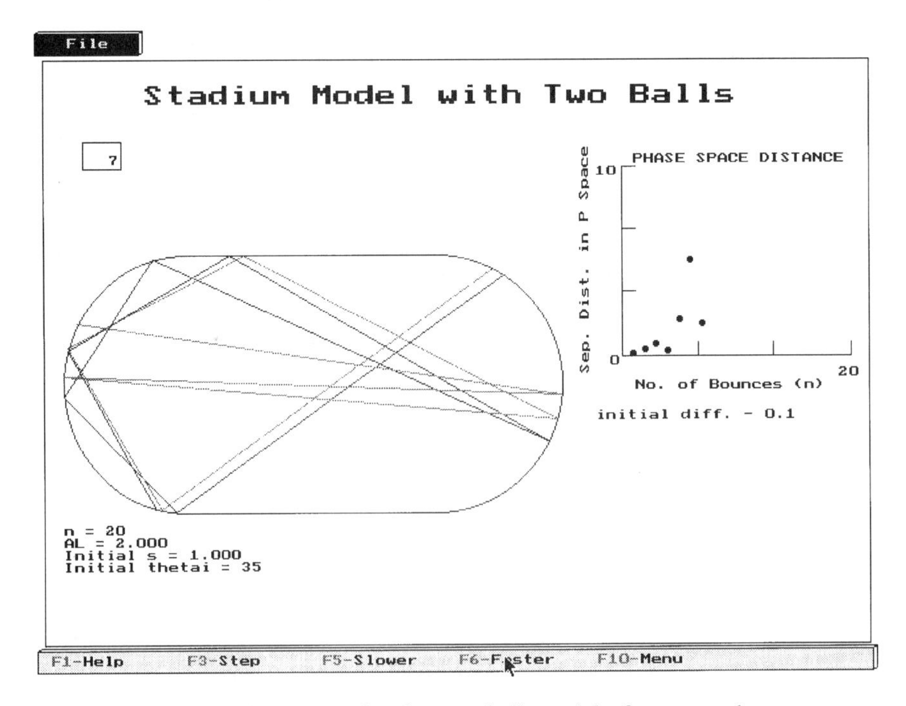

Figure 4.7: The screen for the two-ball model after seven bounces.

4.4 Exercises

An asterisk (*) by an exercise indicates that it is an advanced exercise. Each program creates a disk file that contains data that may be helpful in the analysis of the results. These data files are described in section 4.5.

4.4.1 GALTON

4.1 **Large Number Limit**
Use the traditional Galton board and a large number of balls to examine how well the binomial and Gaussian distributions agree with the final distribution of the balls in the bins.

4.2 **Small Number Limit**
Use the traditional Galton board and a small number of balls to examine how well the binomial and Gaussian distributions agree with the final distribution of the balls in the bins.

4.3 **Custom Galton board***
Try to compute, using the generalized probability distributions discussion found in section 4.2.5, the final distribution of balls in the bins for a custom Galton board. Compare your results to the program's results. Hint: Start with a small Galton board (3 or 4 levels).

4.4 **The Central Limit Theorem***
Use various types of Galton boards (traditional and custom, many levels or few levels, many balls or few balls), to test the central limit theorem. Run GALTON, and select **CLT** in the menu. Use Table 4.1 to determine the goodness of fit of the Gaussian distribution to the results of the 250 samples of the Galton board.†

4.4.2 POISEXP

4.5 **The Poisson Distribution**
Vary the ratio of the half-life to the observation time. Observe the changing shape and expected average of the Poisson distribution. Under what conditions does the Poisson distribution approach the binomial distribution?

4.6 **The Exponential Distribution**
Vary the ratio of the half-life to the observation time.

4.4.3 TWOD

4.7 **Random Walk on a Grid Versus Random Walk on a Plane**
Run a random walk on a grid and the equivalent random walk on a plane. Note that they both produce the same final x-axis distribution (described by a binomial distribution with an expected average of zero), the same

† The test of the central limit theorem takes a long time for large Galton boards, i.e., many levels and/or many balls.

final y-axis distribution (also a binomial distribution with an expected average of zero), and the same expected rms (root mean square) radial displacement, but that the final radial distributions are different. The final radial distribution of a random walk on the plane is described by a Rayleigh distribution, the random walk on the grid is not. Why? The end point of a random walk on a grid is restricted to specific points, while the end point of a random walk on a plane is not.

4.8 **Two-Dimensional Distribution of End Points***
Use the end point distribution data found in the output disk file (TWOD.DAT see section 4.5.5) to plot a three-dimensional contour map with the x displacement plotted on the x-axis, the y displacement plotted on the y-axis, and the number of end points plotted on the z-axis. Is the distribution symmetric about the radius or skewed?

4.9 **Skewed Probability***
Currently the drunk has an equal probability of moving left, right, up, or down. Modify Procedures **Paths** and **InitializeGrid** so that the drunk is more likely to move in one direction than another. Estimate and observe the changes in the distributions.

4.10 **Step Length***
Currently the drunk takes equally spaced steps. Modify Procedures **Paths** and **InitializeGrid** so that the drunk can take variable-length steps. Estimate and observe the changes in the distributions.

4.11 **Restricted Walks***
Modify Procedures **Paths** and **InitializeGrid** to place traps, reflecting walls, and/or absorbing walls on the grid. The walls can be diagonal, vertical, or horizontal. Estimate and observe the changes in the distributions.

4.12 **Self-Avoiding Walks and True Self-Avoiding Walks***
Modify Procedures **Paths** and **InitializeGrid** to create walks in which the drunk cannot visit a site more than once (self-avoiding) or walks in which the probability that the drunk visits a site is a function of the number of times the site has already been visited (true self-avoiding walk). (See reference 6.)

4.13 **Escaping a Boundary***
Rather than stopping the random walk after a specific number of steps, modify Procedures **Paths** and **InitializeGrid** so the random walk stops after the drunk has traveled a specified distance. What is the expected and actual distribution of the number of steps needed to escape? (See ref 4.)

4.4.4 KAC

4.14 **The Small Kac Ring**
Use the small Kac ring to understand the model. In particular, look at the plot of $N(D) - N(L)$, the number of dark balls minus the number of light balls, verses the number of turns. Note the Poincaré cycle and the time reversal features. Also note that, for the small Kac ring, there is

no agreement between the statistical model ($\langle N(D) - N(L)\rangle$ (the average over the canonical ensemble)) and $N(D) - N(L)$. Is this disagreement unexpected?

4.15 **The Large Kac Ring**
The large Kac ring has two parts. In the first part, the Kac ring rotates one half Poincaré cycle from which the average and one sigma deviation of $N(D) - N(L)$ are calculated. In the second part, the Kac ring makes a small number of turns. The number of times $N(D) - N(L)$ exceeds one and two standard deviations (one sigma and two sigmas) of $(N(D) - N(L))$ is counted. Vary the number of markers and choose the number of initial dark balls so that the initial system is in a non-equilibrium condition $(N(D) \neq N(L))$. In the first part, notice how quickly $(N(D) - N(L))$ goes to equilibrium $(N(D) - N(L) = 0)$ and stays there for most of the cycle. Why does $(N(D) \neq N(L))$ leave equilibrium? In the second part, notice that within the approximation to the canonical ensemble (the number of turns is much less than the number of markers), there is good agreement between $(N(D) - N(L))$ and $\langle N(D) - N(L)\rangle$. Why is this agreement expected?

4.4.5 STADIUM

4.16 **Cumulative Computer Computational Errors**
Use the one-ball model to determine how many bounces the ball makes before the computer's finite precision arithmetic lead to round-off errors that affect the resulting numeric answer, based on its failure to retrace its path. The number of bounces that the ball makes before there are cumulative computer computational errors is equal to twice the number of bounces specified in the one-ball model (the ball bounces forward n bounces then back n bounces) and is a function of the stadium's shape and the initial conditions. For periodic trajectories, this number is large and for aperiodic trajectories, it is small. Why?

Look for a relationship between the type of trajectory (periodic or aperiodic) and the Poincaré section (the s versus p plot on the right side of the screen).

4.17 **Periodic and Aperiodic Trajectories**
The stadium's shape and the ball's initial position and momentum determine whether a trajectory is periodic or aperiodic. For some stadium shapes, aperiodic trajectories can be infinitely close to periodic trajectories. For one particular stadium shape, all the trajectories are periodic. Try to determine this stadium shape. Once a periodic trajectory is found, use the two-ball model to determine if the surrounding nearest-neighbor trajectories are periodic or aperiodic.

4.5 Details of the Programs

PROBDRV is the driver program and can be used to call the programs GALTON (the Galton board), POISEXP (the Poisson and exponential distributions), TWOD (random walks in two dimensions), KAC (the Kac ring), and STADIUM (the stadium model). These programs also can be called separately by typing their names.

4.5.1 Input Screen Options for GALTON

The first input screen is used for both traditional and custom Galton boards. It is used to specify the number of balls dropped into the Galton board, the size (number of levels) and type (traditional or customized) of the Galton board, and the probability deflection to the right for a traditional Galton board. The second input screen is used only for customized Galton boards.

The inputs for the first input screen are as follows:

Board type	Traditional Galton board Customized Galton board

For a traditional Galton board, the deflection probability to the right is the same for each pin, and there are no traps, reflecting walls, or absorbing walls. For a customized Galton board, the deflection probability to the right can be defined independently for each pin; and traps, reflecting walls, and absorbing walls can be placed on the board. A trap absorbs balls that come into contact with it. A reflecting wall is a column of pins whose deflection probability to the right is such that any ball hitting one of these pins is directed inward. An absorbing wall is a column of traps.

Number of balls	Integer between 10 and 250 (number of balls dropped)
Number of levels	Integer between 3 and 8 (number of levels of board)

The combination of the number of balls and the number of levels determines whether the Galton board's pins and bins are shown on the screen or just the bins are shown. If the number of levels is greater than six or the number of balls is greater than 100, just the balls dropping into the bins are shown. Otherwise, both the pins and bins are shown.

Deflection probability to right (for traditional Galton board)	Number between 10 and 90 (10 represents a 10% deflection probability; 90 represents a 90% deflection probability)
Output file name	The output disk file name The output file name only appears at the beginning of the program; its default is GALTON.DAT

The second input screen is displayed only for a customized Galton board and displays the pin locations. This screen is used to provide the individual pins' deflection probability to the right, and the locations of the traps, reflecting walls, and absorbing walls. For an individual pin, −1.0 in a deflection probability indicates a trap. Otherwise, the individual pin's deflection probability is a number between 10 and 90.

Place reflecting wall(s) in column(s)	Integer between 1 and 2 × (# of levels −1) (1 represents wall in left-most column; 2 × (# of levels −1) represents wall in rightmost column) Wall can never be placed in center column.
Place absorbing wall(s) in column(s)	Integer between 1 and 2 × (# of levels −1). (1 represents wall in left-most column; 2 × (# of levels −1) represents wall in rightmost column) Wall can never be placed in center column

The columns' numbers are located at the bottom of the screen's sample Galton board.

4.5.2 Output Data File

The output for a Galton Board includes—

- The type of Galton Board (traditional or customized).
- The number of balls dropped.
- The number of levels.
- The deflection probability to the right (one number for a traditional Galton board and the number for each individual pin for a customized Galton board).
- The final bin distribution.
- The corresponding binomial and Gaussian distributions for a traditional Galton board or the predicted distribution for a customized Galton board.
- The actual and theoretical averages.
- The actual and theoretical one sigmas for a traditional Galton board.
- The actual and expected number of balls in the bins for a customized Galton board.

The output for a chi-square goodness of fit test includes—

- The number of bins to which the data (the 250 Galton boards' averages) is assigned.
- The number of Galton board averages in each division (averagecount) and the expected number of Galton board averages in each division assuming a Gaussian distribution.
- The actual and expected average of the 250 Galton boards' averages.
- The chi-square.
- The number of degrees of freedom (the number of bins minus three).

4.5.3 Input Screen Options for POISEXP

Distribution type	Poisson distribution exponential distribution
Half-life	Number between 0.01 and 999 (the half-life in years of a sample in the ensemble)

For the Poisson distribution:

Observation time	Integer between 1 and 60 (minutes that each sample in ensemble is observed)
Number of samples	Integer between 200 and 1,000 (number of samples in ensemble)

For the exponential distribution:

Observation time	Number between 0.001 and 5 (time in half-life units that sample is observed)
Number of evenly spaced observations	Integer between 10 and 30 (Total number of evenly spaced observations made on sample) (Time between observations is observation time divided by number of evenly spaced observations)
Output file name	The output disk file name The output file name only appears at the beginning of the program, its default is POISEXP.DAT

4.5.4 Output Data File

The output for the Poisson distribution includes—

- The half-life in years of a sample in the ensemble.
- The observation time in minutes and years.
- The number of samples in the ensemble.
- The number of particles in a sample.
- The probability that a particle decays during the observation time times the number of particles in a sample.
- A table of the number of decays per observation interval, the number of samples with that number of decays, the corresponding number for a Poisson distribution, and the corresponding number for a binomial distribution.
- The theoretical and the actual averages for the distribution of number of decays.

The output for the exponential distribution includes—

- The number of observations.
- The sample's half-life.
- The decay constant λ (see Eq. 4.45).
- The total years that the sample is observed.
- The years between observations.
- The number of radioactive atoms in the sample.
- A table of the time, the number of radioactive atoms, and the corresponding number of radioactive atoms assuming an exponential distribution.

4.5.5 Input Screens for TWOD

This section describes the input screen options for **TWOD**, random walks in two dimensions.

Type of walk	Random walk confined to two-dimensional grid Random walk confined to a two-dimensional plane
Number of steps (in walk)	Integer between 5 and 10 for random walk confined to grid Integer between 10 and 15 for random walk confined to plane
Number of walks	Integer between 200 and 400
Output file name	The output disk file name The output file name only appears at the beginning of the program, its default is TWOD.DAT

4.5.6 Output Data File

The output includes—

- Whether the random walk is confined to a two-dimensional grid or a two-dimensional plane.
- The number of steps in each walk.
- The number of walks.
- The actual and expected final displacement distribution in the x and y directions.
- The actual and expected average final displacements in the x and y directions.
- The observed value of σ in the x and y directions.
- For a walk confined to a grid, the actual final r-axis displacement and rms radial displacement.
- For a walk confined to a plane, the actual and expected final r-axis displacement and the actual and expected rms radial displacement.
- The distribution of end points.

4.5.7 Input Screen Options for KAC

Ring size	Large ring (ring with 2001 points) Small ring (ring with 9 points)
Number of markers	Integer between 500 and 1500 for large ring Integer between 1 and 8 for small ring
Initial number of dark balls	Integer between 0 and 2001 for large ring Integer between 0 and 9 for small ring
Number of turns the Kac ring makes	For large ring, integer much less than initial number of dark balls; for small ring, 18
Output file name	The output disk file name The output file name only appears at the beginning of the program, its default is KAC.DAT

4.5.8 Output Data File

The output for a small ring is as follows:

- The number of balls (9).
- The number of markers.
- The number of members in the ensemble.
- A table of the number of turns, $N(D) - N(W)$ and $\langle N(D) - N(W) \rangle$.

The output for a large ring is as follows:

- The number of balls (2001).
- The number of markers.
- The initial number of dark balls.
- The number of turns.
- The observed value of $\langle N(D) - N(W) \rangle$.
- The observed value of the standard deviation of $N(D) - N(W)$.
- A table of the number of turns, $N(D) - N(W)$, and $\langle N(D) - N(W) \rangle$.
- The number of points above or below one σ.

- The number of points above or below two σ.
- The average difference between $N(D) - N(W)$ and $\langle N(D) - N(W) \rangle$.
- The standard deviation of the difference between $N(D) - N(W)$ and $\langle N(D) - N(W) \rangle$.

4.5.9 Input Screen Options for STADIUM

Select model	One-ball model The two-ball model
Number of bounces	Number between 1 and 200 (n) For one-ball model, ball bounces n times, then reverses its momentum and bounces n times more For two-ball model, four pairs of non-interacting balls bounce n times, the first pair's initial conditions differ by 0.1, the second pair's by 0.01, ... fourth's by 0.0001
Length of straight portion of stadium (AL)	Number between 0 and 5 0 forms circle, 5 forms long, thin oval (Fig. 4.8)
The initial location of ball (initial s)	Number between 0 and $2(\pi + AL)$ (see Fig. 4.8)
The initial angle θ_i with respect to the x-axis	Number between 0 and 179 degrees, this angle is not the angle of incidence (see Fig.4.8)

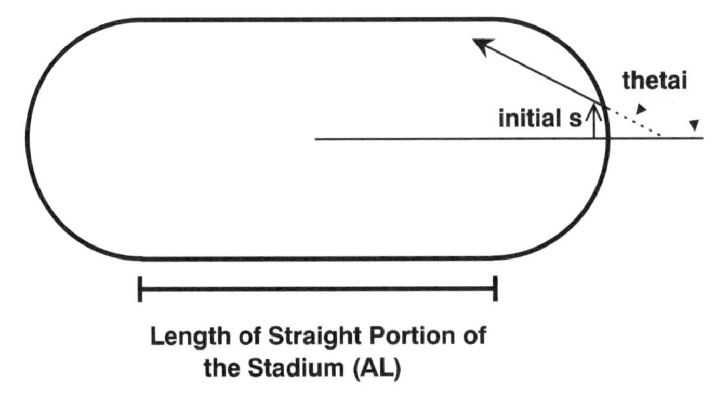

Figure 4.8: Stadium model's initial conditions.

Output file name	The output disk file name The output file name only appears at the beginning of the program, its default is STADIUM.DAT

4.5.10 Output Data File

Each time a stadium model is run, i.e., either the one-ball or two-ball model, the initial conditions are sent to the disk data file. The initial conditions include—

- The length of the straight portion of the stadium (AL).
- The total number of bounces, n.
- The number of balls, nb.
- The stadium's circumference, $2s_{\max}$.
- The ball's initial position on the stadium's circumference, s.
- The initial angle with respect to the x-axis.

For the one-ball model, the following information is provided for each bounce:

- The location where the ball hits the stadium's outline, s.
- The location where the ball hits the stadium's outline with respect to the x, y-coordinates.
- The angle of incidence, α.
- The ball's momentum, i.e., the cosine of the angle of incidence, p.

For the two-ball model, the following information is provided for each bounce:

- The initial nearest-neighbor separation.
- The location where the first ball hits the stadium's outline, $s[1]$.
- The location of where the second ball hits the stadium's outline, $s[2]$.
- The first ball's momentum, $p[1]$.
- The second ball's momentum, $p[2]$.
- The distance in phase space between the first and second ball.

References

1. Barlow, R. *STATISTICS–A Guide to the Use of Statistical Methods in the Physical Sciences.* New York: John Wiley and Sons, 1989.

2. Chandrasekhar, S. Stochastic problems in physics and astronomy. *Reviews of Modern Physics* **15**:1-89, 1943.

3. Gould H., Tobochnik, J. *An Introduction to Computer Simulation Methods–Applications to Physical Systems.* Reading, MA: Addison-Wesley, 1988.

4. Whitney, C. A. *Random Processes in Physical Systems–An Introduction to Probability Based Computer Simulations.* New York: John Wiley and Sons, 1990.

5. Dresden, M. *Studies in Statistical Mechanics.* Amsterdam, The Netherlands: North Holland, 1956, p. 303.

6. Wannier, G. H. *Statistical Physics.* New York: John Wiley and Sons, 1966, p. 387.

7. Gleick, J. *Chaos.* New York, NY: Viking, 1987.

8. Lighthill, Sir J. The recently recognized failure of predictability in Newtonian dynamics. *Proceedings of the Royal Society of London Series.* A **407**:35, 1986.

9. Sperandeo-Mineo, R. M., Falsone, A. Computer simulation of ergodicity and mixing in dynamical systems. *American Journal of Physics* **58**:1073, 1990.

5

Statistical Properties of Gases, Liquids, and Solids

Harvey Gould

5.1 Introduction

Given our knowledge of the laws of physics at the microscopic level, e.g., Newton's laws of motion, how can we understand the behavior of gases, liquids, and solids, and of even more complex macroscopic systems such as polymers and biological systems? One way to help us understand the connections between the microscopic laws of physics and the macroscopic behavior of systems with many degrees of freedom is to use a computer to simulate simple microscopic models of macroscopic systems. Although it is clearly impossible to simulate a typical macroscopic system containing the order of 10^{25} interacting atoms, the simulation of systems with as few as 10–100 particles can give us much useful insight and qualitative information.

Program MANYPART is actually four programs in one and simulates a two-dimensional system of Lennard-Jones particles and hard disks using molecular dynamics and Monte Carlo methods. In sections 5.2–5.4, we discuss the application of constant energy, constant volume molecular dynamics to a system of particles interacting via the Lennard-Jones potential, an archetypical interaction for noble gas atoms. The molecular dynamics of hard disks is discussed in section 5.5. Section 5.6 is devoted to the use of Monte Carlo sampling methods to study the Lennard-Jones system at fixed temperature and density and a system of hard disks at fixed density. The procedure for running the program is discussed in section 5.7. Suggested problems are given in section 5.8 and ideas for extensions of the program are briefly discussed in section 5.9.

5.2 A Simple Model of a Classical Fluid

Consider a system consisting of many identical particles, e.g., argon atoms. Although the most fundamental description of matter at the microscopic level is quantum mechanical, we will assume that we can describe the motion of the individual argon atoms by the laws of classical physics. This assumption is justified if the mean de Broglie wavelength of an atom is much smaller than the mean distance between atoms, a condition that is satisfied if the system is sufficiently dilute or if the temperature of the system is not too low.

Because argon atoms are spherical and chemically inert, we can assume that the force between any two atoms depends only on the distance between them. In this case we can write the total potential energy U as a sum of two-body interactions:

$$U = u(r_{12}) + u(r_{13}) + \cdots u(r_{23}) + \cdots \tag{5.1}$$

$$= \sum_{i<j=1}^{N} u(r_{ij}), \tag{5.2}$$

where the pairwise interaction $u(r_{ij})$ depends only on the magnitude of the separation $\vec{\mathbf{r}}_{ij} = \vec{\mathbf{r}}_i - \vec{\mathbf{r}}_j$ between atoms i and j.

The interaction between atoms is electromagnetic in nature, but involves subtle quantum mechanical effects. In principle, the form of $u(r)$ can be calculated from first principles using quantum mechanics. However, the quantum mechanical calculations of $u(r)$ are very difficult, and it is sufficient to choose relatively simple phenomenological forms for $u(r)$. We know that all molecules have attractive interactions, because all macroscopic systems become liquid at sufficiently low temperatures. For argon and other noble atoms with closed electronic shells, the attractive interaction is due to the mutual polarization of each atom. This attractive interaction is known as the van der Waals interaction and has the form

$$u_{\text{vdw}}(r) \sim -I_0(\sigma/r)^6, \tag{5.3}$$

where I_0 is the ionization potential of the atom, and σ is a measure of the size of the atom.

If two argon atoms approach so closely that their electron shells overlap, the Pauli exclusion principle causes an effective repulsive force that increases rapidly with decreasing separation. The most common phenomenological form of the potential energy that includes both repulsive and attractive effects is known as the Lennard-Jones potential and is given by

$$u(r) = 4\epsilon\left[\left(\frac{\sigma}{r}\right)^{12} - \left(\frac{\sigma}{r}\right)^{6}\right]. \tag{5.4}$$

The form of the repulsive part of the potential, r^{-12}, is chosen for convenience and historical reasons only. Note that the Lennard-Jones potential is characterized by a length σ and an energy ϵ. The properties of $u(r)$ in Eq. 5.4 are studied in Exercise 5.1. The values of ϵ and σ for argon are $\epsilon/k = 119.8$ K and $\sigma =$

3.405 Å. The quantity k is Boltzmann's constant. The mass m of an argon atom is 6.69×10^{-26} kg.

Because it is not convenient to use very small or very big numbers in a computer program, we will choose units such that the length, energy, and mass are measured in units of σ, ϵ, and m, respectively. Hence we measure velocities in units of $(\epsilon/m)^{1/2}$, time in units of $\tau = (m\sigma^2/\epsilon)^{1/2}$, and the pressure in terms of ϵ/σ^2 (see Exercise 5.2).

5.3 Boundary Conditions and the Choice of Ensemble

Imagine a gas of N particles in a box of volume V. We will consider two-dimensional systems because they are easier to visualize on the screen. In this case the volume of the box is an area, but we will refer to it as a volume. The simplest physical situation corresponds to the entire system (box plus particles) being in isolation, i.e., the particles interact only among themselves and with the walls. We are interested in the thermodynamic limit, $N \longrightarrow \infty$ and $V \longrightarrow \infty$, such that the density $\rho = N/V$ is a constant. In the large V limit, the shape of the box is irrelevant for a liquid, so we will choose a square box with linear dimension L so that $V = L^d$, where d is the spatial dimension. The relative number of particles near the walls of the box is proportional to $N^{(d-1)/d}/N = N^{-1/d}$ (see Exercise 5.3), and hence the relative number of particles near the walls goes to zero as $N \longrightarrow \infty$. Because the number of particles that can be simulated on a computer is the order of 10^2–10^6, the fraction of particles near the walls would not be small for such systems. Hence we cannot do a realistic simulation of a macroscopic system by allowing the particles to interact with a wall, e.g., by assuming that the particles reflect off a rigid wall.

The easiest way to minimize surface effects and to simulate more closely a macroscopic system is to use periodic boundary conditions. In brief, the idea is to replace the box with its four ($d = 2$) or six ($d = 3$) walls by a cell with imaginary surfaces. If a particle crosses a surface of the cell, it reenters the cell through the opposite surface with an unchanged velocity. In this way we keep the system at constant density, but eliminate the rigid walls and represent a macroscopic system more closely.

The use of periodic boundary conditions implies that the central cell is repeated an infinite number of times; i.e., the small volume of the central simulation cell is part of an infinite system. Hence if particle i is at $\mathbf{r}_i$, there are image particles at $\mathbf{r}_i + \mathbf{n}L$, where $\mathbf{n} = (n_x, n_y)$ and n_x and n_y are integers (see Fig. 5.1). In general, there are an infinite number of contributions to the force on any given particle. However, for short range interactions, we may reduce the number of contributions by assuming that the separation r_{ij} between particle i at $\mathbf{r}_i$ and particle j at $\mathbf{r}_j$ is $r_{ij} = \min |\mathbf{r}_i - \mathbf{r}_j + \mathbf{n}L|$ for all $\mathbf{n}$. That is, a particle in the center cell interacts with each particle in the center cell or its nearest image, and the maximum distance between any two particles in the x or y directions is $L/2$. This adaptation of periodic boundary conditions is known as the nearest (minimum) image convention.

Because the particles are kept at constant density by the boundary conditions, the system of particles has a fixed total energy E and fixed values of N and V. These conditions are equivalent to the microcanonical ensemble in statistical mechanics.

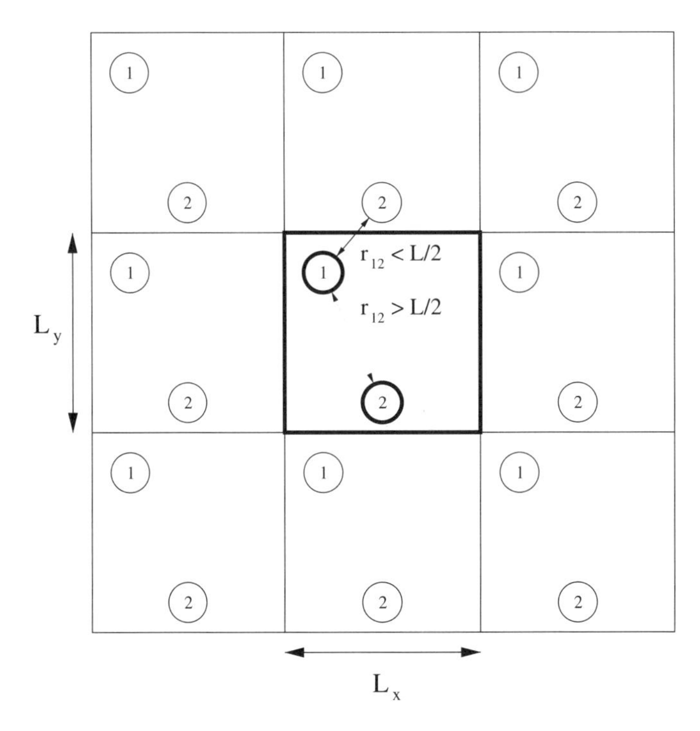

Figure 5.1: The simulation cell and its periodic images. The center cell is replicated to form an infinite lattice.

We will add the additional condition that the total momentum of the center of mass of the particles is zero; i.e., the box is fixed. We discuss the simulation of other statistical ensembles in section 5.6.

5.4 Molecular Dynamics

Given the interaction potential between any two particles and the boundary conditions, our major task is to solve Newton's equations of motion for each particle. That is, we need to solve

$$\frac{d\mathbf{r}_i}{dt} = \frac{\mathbf{p}_i}{m} \tag{5.5}$$

$$\frac{d\mathbf{p}_i}{dt} = \sum_{j \neq i}^{N} \mathbf{F}(r_{ij}), \tag{5.6}$$

where $\mathbf{F}(r_{ij})$ is the force on particle i due to particle j. The equations of motion Eqs. 5.5 and 5.6 cannot be solved analytically in general; we will solve them numerically by converting them to finite difference equations. For simplicity, we

there is no attractive interaction, we might expect that there is no transition from a gas to a liquid. Is there a phase transition between the dense fluid and a solid? What is the difference, if any, between the properties of a dense liquid of particles interacting via the Lennard-Jones potential Eq. 5.4 and a system of hard spheres or disks?

The numerical methods for integrating Newton's equations of motion for a continuously varying potential such as Eq. 5.4 are not applicable to the discontinuous hard sphere interaction Eq. 5.15. The reason is that whenever the distance between the particles equals σ, the particle velocities change instantaneously. Instead of solving the equations for finite time steps, we have to move the particles in straight lines until a collision occurs. The general procedure can be summarized as follows.

1. For each pair of particles i, j, compute the time of their collision, t_{ij}, ignoring the presence of other particles. (The collision time will be infinite for some pairs).
2. Find the minimum collision time.
3. Move all the particles in the direction of their velocities until the collision occurs.
4. Change the velocities of the colliding pair.
5. Compute the physical properties of interest and repeat steps 1–4.

To find the collision, consider disks i and j with positions $\mathbf{r}_i(0)$ and $\mathbf{r}_j(0)$ and velocities $\mathbf{v}_i(0)$ and $\mathbf{v}_j(0)$ at time $t = 0$. If these particles are to collide at time t_{ij}, the following relation must be satisfied:

$$|\mathbf{r}_i(t_{ij}) - \mathbf{r}_j(t_{ij})| = \sigma \tag{5.16}$$

$$\mathbf{r}_i = \mathbf{r}_i(0) + \mathbf{v}_i(0)t_{ij} \tag{5.17}$$

$$\mathbf{r}_j = \mathbf{r}_j(0) + \mathbf{v}_j(0)t_{ij}\,. \tag{5.18}$$

Hence

$$|\mathbf{r}_{ij} + \mathbf{v}_{ij}t_{ij}|^2 = \sigma^2, \tag{5.19}$$

where $\mathbf{r}_{ij} = \mathbf{r}_i - \mathbf{r}_j$ and $\mathbf{v}_{ij} = \mathbf{v}_i - \mathbf{v}_j$. If we let $b_{ij} = \mathbf{r}_{ij} \cdot \mathbf{v}_{ij}$, then Eq. 5.19 can be written as a quadratic equation:

$$t_{ij}^2 v_{ij}^2 + 2b_{ij}t_{ij} + r_{ij}^2 - \sigma^2 = 0\,. \tag{5.20}$$

If $b_{ij} > 0$, then the disks are going away from each other and will not collide. If $b_{ij} < 0$, then Eq. 5.20 might still have complex roots in which case no collision occurs. Otherwise there are two positive roots, the smallest of which corresponds to the time of collision:

$$t_{ij} = \frac{-b_{ij} - \left[b_{ij}^2 - v_{ij}^2(r_{ij}^2 - \sigma^2)\right]^{1/2}}{v_{ij}^2}. \tag{5.21}$$

After we have found all the collision times for all possible collisions and saved them in an array, we have to find the minimum collision time, t_{ij}, and the colliding pair i and j. Then all particles are moved in a straight line for the time t_{ij}, periodic boundary conditions are applied, and all other collision times are reduced by the time t_{ij}.

We next need to change the velocities of the pair of colliding particles. If we apply conservation of linear momentum and kinetic energy and assume equal masses, we find that

$$\mathbf{v}_i(\text{after}) = \mathbf{v}_i(\text{before}) + \Delta\mathbf{v} \tag{5.22}$$

$$\mathbf{v}_j(\text{after}) = \mathbf{v}_j(\text{before}) - \Delta\mathbf{v}, \tag{5.23}$$

where the change in the velocity $\Delta\mathbf{v}$ is given by

$$\Delta\mathbf{v} = -(b_{ij}/\sigma^2)\mathbf{r}_{ij}, \tag{5.24}$$

and $b_{ij} = \mathbf{r}_{ij} \cdot \mathbf{v}_{ij}$ is computed at the time of impact. It is not necessary to recompute all the collision times and collision partners all over again. Instead we can simply update the collision partners of i and j and any other particles that were due to collide with i and j.

If you think about the nature of a system of hard disks, you will realize that the speeds of the particles only set the time scale. For this reason the temperature of a system of hard disks is not really relevant, even though we can compute it. Of course the mean pressure is relevant, and it is convenient to write Eq. 5.12 in a form involving an average over collisions:

$$PV = NkT + \frac{1}{d}\frac{1}{t} \sum_{\text{collisions}} \Delta\mathbf{p}_i \cdot \mathbf{r}_{ij}, \tag{5.25}$$

where i and j are the colliding pair, $\mathbf{r}_{ij}$ is the vector between i and j at the time of the collision, and $\Delta\mathbf{p}_i = -\Delta\mathbf{p}_j$ is the change in their momentum. The average is over all collisions in the observation time t.

Are there any other physical quantities that are important for hard disks? Because geometry is important for hard disks, we might imagine that the geometry or the structure of the system also is interesting. For example, if particle i is at $\mathbf{r}_i$, what is the probability that another particle is a displacement $\mathbf{r}$ away? The simplest measure of the structure of a fluid is the pair distribution function $g(r)$, which can be measured by x-ray scattering in laboratory systems. The function $g(r)$ gives the probability of finding a pair of particles a distance r apart, relative to the probability of a completely random distribution at the same density. The procedure is to choose a particle as the origin and to compute the number of particles that are in the interval r and $r + dr$ from the origin. This histogram equals $\rho g(r)\, d\mathbf{r}$, where the volume element $d\mathbf{r}$ is $4\pi r^2 dr$ in $d = 3$ and $2\pi r dr$ in $d = 2$. The histogram is averaged over all possible particles.

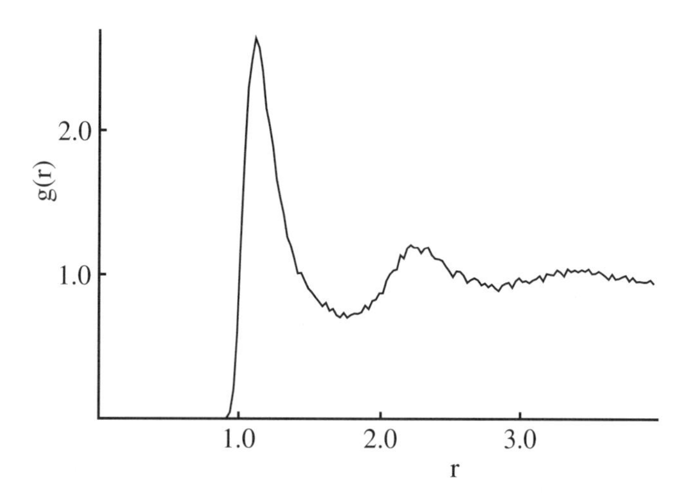

Figure 5.2: The pair correlation function $g(r)$ for a two-dimensional Lennard-Jones system of $N = 36$ particles at a density $\rho = 0.50$ and $T = 0.40$ averaged over a time of $t = 10\,\tau$. The results for $g(r)$ would be smoother if a larger system was sampled for the same time interval or if a much longer time interval was used for the same size system. Note the peaks in $g(r)$ at $r \approx 1$ and $r \approx 2.2$. More relative maxima would be observed at higher densities for larger systems.

The determination of the pair correlation function is of interest for any system, not just for hard spheres or hard disks. In Figure 5.2 we show the typical results for the dependence of $g(r)$ on r for a dense fluid.

5.6 Monte Carlo Methods

So far we have discussed the computation of various thermodynamic quantities using time averages, i.e., averages over the trajectory of the system in phase space. However, the use of ensemble averages is much more useful in theoretical calculations. The quasi-ergodic hypothesis asserts that the time average and the ensemble average are equivalent if the quantities which are held constant are the same.

Let us first consider how we might do an ensemble average for a system of hard disks. In this case we can specify the ensemble by the values of N and V; i.e., we consider fixed density $\rho = N/V$. For this ensemble we would like to generate an arbitrary number of systems with the same values of N and V, but with different configurations of hard disks. The method we will use is an example of a *Monte Carlo* method; i.e., it is based on the use of a sequence of random numbers to generate an unbiased sample of configurations.

Suppose that we have an acceptable configuration of hard disks, i.e., a configuration in which no two hard disks overlap. Our problem is to generate other acceptable configurations that are statistically independent of the first. One way to generate a new configuration without bias is to close our eyes and move a particular disk at random. We can implement this procedure on a computer as follows. Choose a disk at random and move it to a new trial position. If the trial position of the disk overlaps another disk, the move is rejected and the old configuration is retained and recounted in the calculation of averages; otherwise, the move is accepted. The only variable is the choice of the parameter δ, the maximum displacement. If δ is too small, most moves will be accepted, but the configurations will be strongly correlated and the generation of statistically independent configurations would be inefficient. If δ is too large, it is likely that most moves will be rejected. We have chosen δ to be 20% of the mean interparticle separation. The exploration of the optimum choice of δ is explored in Exercise 5.18.

Because hard disks do not actually come into contact using the Monte Carlo method, we cannot use Eq. 5.12 to compute the mean pressure. Instead we can express the virial in terms of $g(r)$ at contact:

$$\frac{\beta P}{\rho} = 1 + \frac{\pi}{2}\rho\sigma^2 g(r = \sigma^+), \qquad \text{(two dimensions)} \tag{5.26}$$

where $\beta = 1/kT$, and $g(r = \sigma^+)$ is the value of the pair correlation function at contact. The best way to calculate $g(r = \sigma^+)$ is to compute $g(r)$ for different values of r and then to extrapolate the results to $r = \sigma$.

We next consider the simulation of a system of Lennard-Jones particles in the canonical ensemble where N, V, and T (rather than the total energy E) are fixed. That is, we want to generate configurations of a system in equilibrium with a heat bath at temperature T. In this ensemble the configurations are not distributed with equal probability, but occur with the Boltzmann probability:

$$P_n = \frac{1}{Z} e^{-\beta E_n}, \tag{5.27}$$

where n labels the microstate. Because $\sum_n P_n = 1$, the normalization constant Z in Eq. 5.27 is given by

$$Z = \sum_n e^{-\beta E_n}. \tag{5.28}$$

The summation in Eq. 5.28 is over all the possible microstates of the system.

How can we generate configurations with the correct Boltzmann probability Eq. 5.27 in the canonical ensemble? One way to formulate the procedure is in terms of random walks and importance sampling,[2] but we will give a plausibility argument instead. Suppose that we have a configuration n with energy E_n, and we want to know the probability of a transition to microstate m with energy E_m. Because the latter configuration occurs with probability $P_m = \exp(-\beta E_m)/Z$, the ratio of the probabilities is given by $P_m/P_n = \exp(-\beta(E_m - E_n) = \exp(-\beta\Delta E)$, where $\Delta E = E_m - E_n$ is the change in the energy of the system. The principle of detailed balance[5] requires that the ratio of transition probabilities satisfies the condition

$$W(n \longrightarrow m)/W(m \longrightarrow n) = e^{-\beta(E_m - E_n)}, \tag{5.29}$$

where $W(n \longrightarrow m)$ is the probability that the system makes a transition from microstate n to m. One way of satisfying this condition is to choose the transition probability of going from state n to state m to be unity if $\Delta E \leq 0$, and equal to $\exp(-\beta\Delta E)$ if $\Delta E > 0$.

The Monte Carlo procedure with this transition probability can be summarized as follows. For a given configuration of particles, choose a particle at random, and displace it in a random direction with maximum displacement δ. If the change in the total energy ΔE is less than or equal to zero, accept the trial. Otherwise compute the quantity $w = \exp(-\beta\Delta E)$ and generate a uniform random number r in the interval 0 to 1. If $r \leq w$, accept the trial move; otherwise retain the old configuration. Compute the values of the desired physical quantities after N trial moves. These steps are repeated until a sufficient number of independent configurations is obtained.

It is straightforward to compute the mean energy $\langle E \rangle$ and the mean squared energy $\langle E^2 \rangle$ as well as the pair correlation function $g(r)$ as a function of the temperature T of the heat bath. In the thermodynamic limit, the relation $T(E)$ obtained in the microcanonical ensemble and the relation $\langle E(T) \rangle$ obtained in the canonical ensemble give equivalent results. However, the results are not identical for finite systems, especially for the relatively small systems that we will use. And even though the mean values of various quantities are identical in the thermodynamic limit, the fluctuations of these quantities are not the same. For example, the total energy does not fluctuate in the microcanonical ensemble, but fluctuates in the canonical ensemble. It can be shown that the energy fluctuations in the canonical ensemble are related to the heat capacity C_V by[5]

$$C_V = \frac{1}{kT^2}(\langle E^2 \rangle - \langle E \rangle^2). \tag{5.30}$$

5.7 Procedure for Running Program MANYPART

The default screen and choices correspond to a molecular dynamics simulation of a fluid of $N = 16$ particles interacting via the Lennard-Jones potential. The total energy and the initial values of the temperature and pressure are shown in the right-bottom window. If the default choices are acceptable, the simulation can be started by either clicking the mouse in the trajectory window or by pressing hot key F2. Help can be obtained at any time by pressing hot key F1.

In the molecular dynamics simulation the pressure and temperature are plotted as a function of the (dimensionless) time in the upper left and right windows, respectively. Plots in yellow represent running averages and plots in green represent the value of the quantity at time t. The pressure is computed from the virial (see Eq. 5.12). The calculation of running averages can be reset by pressing hot key F4. The plot in the upper-left window can be switched between the pressure and the pair correlation function $g(r)$ by clicking on the upper-left window; the plot in the upper-right window can be switched between the temperature and the mean square displacement $\langle R^2 \rangle$ in an analogous way. The speed distribution $P(v)$ and the velocity distribution $P(v_x)$ can be seen in the middle-right window. The positions of the particles are seen in the lower-left window. Because the program will run faster if the trajectories are not drawn on the screen, it is convenient to use hot key F3 to toggle between showing and hiding this window. The trajectories can be drawn anew by pressing hot key F5.

The menu headings allow the user to exit the program, change the simulation algorithm, and modify various settings. Under **File**, the user can read **About the Program, About CUPS**, and about the choice of **Units**. A new configuration of particle positions and velocities can be chosen by selecting **New**, a previously saved configuration can be used by choosing **Open**, and an existing configuration can be saved in a file by selecting **Save**. The Lennard-Jones and hard disk interactions can be chosen under the **Molecular Dynamics** and **Monte Carlo** menu headings. Under **Settings** the user can make changes in the **Temperature**, **Time Step** (Δt), and the **Box Length** (L_x and L_y), and in some of the output parameters.

The menu structure for the other algorithms and interactions is similar. Note that the temperature is not plotted for the molecular dynamics simulation of hard disks since the kinetic energy is a constant. The temperature of the heat bath in the Monte Carlo simulation of the Lennard-Jones potential is set by a slider.

5.8 Exercises

The molecular dynamics and Monte Carlo methods incorporated in the program are general-purpose tools for simulating classical many-particle systems. Thus Program MANYPART can be applied in many ways that are not mentioned in the exercises or in the text. See, for example, the texts by Allen and Tildesley,[2] Gould and Tobochnik,[4] Haile,[3] and Heermann[6] for much more complete discussions of molecular dynamics and Monte Carlo methods.

5.1 **Lennard-Jones Interaction**
Sketch the r-dependence of the Lennard-Jones potential Eq. 5.4 and the corresponding force $f(r) = -du/dr$. Find the value of r at which the Lennard-Jones potential has a minimum. At what value of r does the force change sign?

5.2 **Units**
Show that the appropriate combination of mass m, energy ϵ, and length σ with dimensions of pressure in two dimensions is ϵ/σ^2. What is the appropriate combination of these parameters with the dimensions of time?

5.3 **Surface Particles**
Show that the number of particles near the walls of a three-dimensional box is proportional to $N^{2/3}$, and hence the fraction of particles near the walls is proportional to $N^{-1/3}$. Generalize this result to d dimensions. Hint: Assume the box is spherical.

5.4 **Approach to Equilibrium**
Use the molecular dynamics algorithm in Program MANYPART to simulate a system of $N = 16$ hard disks in a square box with linear dimension $L = 5$. (Remember that all lengths are measured in terms of σ.) Let the system run for a while. What do you observe? Then use the **Settings** menu to expand the box in the horizontal direction so that $L_x = 12$. Is the system initially in equilibrium? How would you describe the evolution of the system? Does the system tend to fill the box uniformly? Does the system approach equilibrium? You can make your observations more quantitative by counting the number of particles on the left-half of the box. The easiest way to do so is to use the F2 hot key to pause the system and count the number of particles visually. If time permits, consider larger systems, e.g., $N = 36$ and $N = 64$. Be sure to make the initial simulation cell sufficiently large. Is the notion of equilibrium better defined for larger systems? Is the

distribution of particle positions more-or-less uniform in equilibrium if the number of particles is sufficiently large? You can explore similar questions with the Lennard-Jones potential.

5.5 **Special Initial Conditions**

We found in Exercise 5.4 that one of the characteristic features of macroscopic systems is that an isolated system in a nonuniform state will change over time from a less random to a more random state. However, suppose we start with all the particles in a specially prepared state where they all are at rest and uniformly spaced in the vertical direction. Does the system still approach equilibrium? Choose **Open** from the **File** menu and open the file SPEC11A.DAT. Start the molecular dynamics simulation using either hard disks or the Lennard-Jones interaction. What happens? Next, suppose we change the position of the center particle by 1 percent in the upward direction. What do you think happens now? Open the file SPEC11B.DAT and run the simulation. Is it still correct to state that macroscopic systems approach random configurations?

5.6 **Sensitivity to Initial Conditions**

Begin with the initial configuration SPEC16.DAT and save the configuration after you have run for a short time t (or number of collisions). Then edit the file SPEC16.DAT and change the position of one particle 0.1 percent. Run the dynamics for the same amount of time and compare the configurations. Are the particle positions and velocities more-or-less the same? Is the system sensitive to its initial conditions?

5.7 **Time Reversal**

a. Begin with the initial configuration SPEC16.DAT and save the positions and velocities of the particles at regular time intervals (Lennard-Jones) or number of collisions (hard disks). (Choose **Save** from the **File** menu.) Then consider the time-reversed process, $t \longrightarrow -t$, which would occur if the velocities of all the particles were reversed. The easiest way to change the velocities is to edit the File and to change the signs directly. Then choose **Open** from the **File** menu and start the simulation from the time-reversed final configuration. Do the particles return to their original initial condition?

b. Repeat the same procedure, but do not reverse the velocities until twice the time (or number of collisions) has elapsed. How long can you let the system run before the time-reversed system does not return to its original initial condition? Explain your results.

5.8 **Averages Over Trajectories**

Most texts on statistical physics emphasize that probabilistic concepts are needed to describe systems with many degrees of freedom because of the overabundance of information. However, on the basis of your results in Exercises 5.5–5.7, explain why we need to use probabilistic concepts to describe even a relatively small number of particles. Are the computed trajectories of the particles meaningful? Are averages over the computed trajectories meaningful?

5.9 **Conservation of Total Energy**

Because the system is isolated, the total energy should be conserved exactly in the molecular dynamics simulation of a continuous potential. (We say that E is a constant of the motion.) However, because we have replaced the differential equations of motion by finite difference equations, the energy is no longer conserved exactly. The idea is to choose the time step Δt small enough so that E is conserved to the desired precision. Modify the molecular dynamics part of the program so that the total energy E at each time step is saved in a file. Or pause the program periodically and record the value of E. How well is the total energy conserved? Does E drift upward or downward with time or is it bounded? Choose **time step** from the **Settings** menu and confirm qualitatively that E is better conserved when Δt is made smaller. The numerical algorithm used in the text is said to be second-order, namely, the fluctuations in E should be proportional to $(\Delta t)^2$, if Δt is sufficiently small. Compute $(\Delta E)^2 = \overline{E^2} - \bar{E}^2$, where the average is over at least 100 time steps, and confirm that the program is working correctly. The default value of Δt in the program is $\Delta t = 0.01$ (in dimensionless units), but its optimum value depends on the density and the mean temperature.

5.10 **Distribution of Velocities**

Run the molecular dynamics part of the program using either the Lennard-Jones potential or hard disks. Choose either the initial default configuration or one of the configurations on the disk, e.g., M32.DAT or M64.DAT. Run the program long enough to confirm that the system is in equilibrium. What criteria did you use for equilibrium? What is the qualitative behavior of the speed distribution $P(v)$ and the velocity distribution $P(v_x)$ on v and v_x, respectively? What is the most probable value of v and v_x? What is the approximate width of these distributions? Change the temperature from the **Settings** menu and determine how $P(v)$ and $P(v_x)$ depend on temperature. Do these distributions depend on density (at fixed T?) (The bin size for computing these distributions is $\Delta v = 0.1$.)

5.11 **Qualitative Properties of a Solid**

Choose **New** from the **File** menu and start the system from a square lattice. Choose $N = 16$, $L = 4.2$, and $T^* = 0.25$. Run the system for a sufficiently long time to determine if the system is a solid. What is the behavior of $\langle R^2(t)\rangle$? Describe the qualitative nature of the positions of the particles. What are your criteria for a solid? After the system has reached equilibrium, increase the temperature by 20% by rescaling all the velocities from the **Settings** menu. Continue this process until you have enough data to plot C_V as a function of T. What is the qualitative dependence of C_V on T? What happens if you continue increasing T?

5.12 **Generation of Initial Configurations for Molecular Dynamics**

The generation of initial configurations for the Lennard-Jones potential that are representative of configurations at the desired values of T and ρ is a time-consuming job in molecular dynamics. One problem is that we can control the total energy E, but not T. From Exercise 5.11 we can see that one way of generating an initial configuration of particles at the desired density and temperature is to place the particles on the sites of a regular lattice and give them random velocities. Then gradually raise the mean square velocity until the desired temperature is reached. What happens if the velocities are increased too quickly? What is the virtue of starting from a lattice? Another way of generating initial conditions is to do a Monte Carlo simulation by placing the particles at random. When the system has reached equilibrium, the particle velocities can be chosen at random from the Maxwell-Boltzmann velocity distribution.

5.13 **Structure of a Dense Liquid and a Solid**
Measure the mean pressure and pair distribution function $g(r)$ for a system of $N = 36$ hard disks at $\rho^* = 0.88$ and $T^* \sim 1.0$. Explain the qualitative r-dependence of $g(r)$. Then use one of the hard disk configurations to simulate a system of Lennard-Jones particles at the same value of ρ and T. Compare the values of P and $g(r)$ for the two systems. Does the repulsive or the attractive part of the Lennard-Jones potential play a more important role in determining the structure of a dense liquid?

5.14 **Mean Free Time for Hard Disks**
The mean free time t_c is the average time a particle travels between collisions. Suppose that we know that 32 collisions occurred in a time $t = 1.2$ for a system of $N = 16$ hard disks. Because two disks are involved in each collision, there were an average of $(2 \times 32)/16$ collisions per particle. Hence $t_c = 1.2/(64/16) = 0.3$. How does t_c depend on the temperature for fixed density? Is the temperature a relevant quantity? How does t_c depend on the density?

5.15 **Qualitative Behavior of the Self-Diffusion Coefficient**
Use one of the equilibrium fluid configurations that you generated in Exercise 5.13 and determine how the mean square displacement $\langle R^2(t) \rangle$ increases with t. Does $\langle R^2(t) \rangle$ increase as t^2 as it would for a collection of free particles or approximately linearly with t? If $\langle R^2(t) \rangle$ does increase linearly, estimate the slope and use Eq. 5.14 to estimate D. Obtain D for several temperatures and give a simple explanation for the qualitative dependence of D on T that you observed. Develop an analogous explanation for the dependence of D on ρ. Then compute $\langle R^2(t) \rangle$ for an equilibrium configuration corresponding to a harmonic solid. What is the qualitative behavior of $\langle R^2(t) \rangle$ in this case? If time permits, determine D for several densities for a hard disk system and qualitatively compare the density-dependence of D and t_c (see Exercise 5.14.)

5.16 **Monte Carlo Configurations**
The goal of Monte Carlo simulations of a fluid or solid is to generate statistically independent configurations that are distributed with the correct probability. Why are only the positions of the particles changed and the particle velocities irrelevant?

5.17 **Symmetry of a Two-Dimensional Lennard-Jones or Hard Disk Solid at Low Temperature**
Begin with a square lattice of $N = 100$ Lennard-Jones or hard disk particles in a central cell with $L = 11$ at $T = 0.5$. How many nearest neighbors does

each particle have in a perfect square lattice? Does the symmetry of a square lattice correspond to the symmetry with the minimum energy? Although we could compute the energy of different configurations, it is interesting to see if the dynamics will lead the system to "find" the desired symmetry. Run the program for at least $t = 2\tau$ or 1000 collisions and determine how many nearest neighbors most particles have.

5.18 **Optimum Maximum Displacement**
The optimum choice of the maximum displacement δ in a Monte Carlo simulation of a system of particles is not known in general. As mentioned in the text, if δ is too small, then most moves will be accepted, but the configurations will be strongly correlated and the generation of statistically independent configurations would be inefficient. And if δ is too large, most moves will be rejected. One rule of thumb is that δ should be chosen so that approximately 50% of the trial moves are accepted. Another criterion is that δ should be chosen to maximize the mean square displacement $\langle R^2(t)\rangle$, and hence the rate of exploration of different configurations. Note that although the purpose of a Monte Carlo method is to generate equilibrium configurations, we can interpret the transition from one configuration to another as a type of pseudodynamics. In this context one unit of time is equivalent to N trial moves, that is, on the average each particle has an equal chance to move. (This definition of time has no obvious relation to the time in a molecular dynamics simulation.) Find the value of δ that maximizes D for a given value of T and ρ. What is the acceptance ratio in this case?

5.9 Program Modifications

The easiest change to make is to use other interactions besides the Lennard-Jones potential. Examples of interactions that are of current interest include the model potentials:

$$u(r) = (u_0/r)e^{-r/\sigma} \qquad \text{Debye or Yukawa}$$

$$u(r) = u_0 e^{-r^2/\sigma^2} \qquad \text{Gaussian}$$

$$u(r) = u_0/r^n, \qquad \text{inverse power,}$$

where u_0 is a constant. The potential and force are specified in Procedures **U** and **f** respectively. The most common choices of n are $n = 12, 9, 6$, and $n = 1$. The choice $n = 1$ corresponds to a long-range interaction and cannot be implemented without extensive modifications to the program. In particular, the nearest (minimum) image convention (see section 5.3) must be replaced by a sum over all the images of the particles in the infinite number of periodic cells. It also is possible to use more realistic interactions including three-body forces.

Another artificial limitation of the program is that it is restricted to a maximum of 200 particles. This restriction can be removed by changing the value of **Nmax**, but the program will run slowly if N becomes too large. The present program becomes inefficient for sufficiently large N because the force on each particle is recomputed at each time step by determining the distance to all the other $N - 1$ particles. Such a time-consuming calculation is not necessary for short-range forces because we can neglect the force due to those particles that are more than a certain distance away. For example, the force at $r = 3\sigma$ for the Lennard-Jones interaction is approximately three orders of magnitude smaller than it is at $r = \sigma$. Hence, for short-range interactions we can save time by only computing the force due to particles that are within the desired cutoff distance. Of course, we have to know these particles in advance or we would not save any time. One procedure is to save the particles that are within range in an array and to update this array at regular time intervals. The necessary modifications (and complications) do not become efficient until the number of particles is of order 10^3.

The molecular dynamics part of the program can be modified to simulate other ensembles, in particular, the constant N, V, T ensemble and the constant N, P, H ensemble (H is the enthalpy), rather than the constant N, V, E ensemble used in the program. One way to implement constant pressure molecular dynamics is to couple the system to an external variable V, the volume of the simulation cell. The idea of the coupling is to mimic the action of a piston and to vary the volume so as to keep the instantaneous pressure close to the desired pressure. The piston is given a fictitious mass and the equations of motion are generalized so that the differential equation of motion for the volume of the box is coupled to the equations of motion for the particles. A method for implementing constant temperature molecular dynamics is based on a similar coupling between the particles in the simulation cell and an extra degree of freedom that represents the heat bath.

Transport coefficients such as the thermal conductivity and the shear viscosity are of much interest. The easiest way conceptually to compute these quantities is to determine the response of the system to a small perturbation. For example, we might imagine computing the viscosity by applying a shear force to the system. However, the calculation of accurate results for the transport coefficients requires experience with computer simulations, larger systems, and long run times.[2,3,4] However, if you wish to do simulations on systems of more than several hundred particles, it is strongly recommended that you write a program for that purpose.

References

1. Reif, F. *Statistical Physics.* New York: McGraw-Hill, 1967.

2. Allen, M. P., Tildesley, D. J. *Computer Simulation of Liquids.* New York: Oxford University Press, 1987.

3. Haile, J. M. *Molecular Dynamics Simulation.* New York: John Wiley and Sons, 1992.

4. Gould, H., Tobochnik, J. *An Introduction to Computer Simulation Methods.* 2nd ed. Reading, MA: Addison-Wesley, 1995.

5. Reif, F. *Fundamentals of Statistical and Thermal Physics.* New York: McGraw-Hill, 1965.

6. Heermann, D. W. *Computer Simulation Methods.* 2nd ed. New York: Springer-Verlag, 1990.

6

Quantum Ideal Gas

Jan Tobochnik

6.1 Introduction

This chapter discusses the background needed to understand the two programs on quantum ideal gases. Program QMGAS1 calculates various quantities using the grand canonical ensemble, and program QMGAS2 does a Monte Carlo simulation in the canonical ensemble. The two programs are discussed in section 6.3 and section 6.4, respectively.

Macroscopic systems involve large numbers of interacting particles. In most cases the exact calculation of thermodynamic properties using statistical mechanics is impossible because of the mathematical difficulties involved. However, there is an important class of systems for which the calculations can be done exactly and to a large degree analytically.[1–3] These systems are those for which the total energy is the sum of individual single particle energies or for which the dynamical degrees of freedom can be transformed to normal modes such that the total energy is the sum of the individual energies associated with each mode. Although there is no potential energy of interaction between particles or modes, a particle can change its energy state by exchanging energy with a heat bath.

There are many systems that are approximately ideal. For example, there are dilute monotonic gases where the energy for each particle is almost totally kinetic. Crystals provide another example. Here the atoms are strongly coupled by spring-like forces between nearest neighbors, but the equations of motion can be reduced to a system of noninteracting normal modes that represent the fundamental modes of vibration or sound waves. In the quantum version of this model the sound waves are quantized and their quanta of energy are called phonons. Blackbody radiation, which can be represented as an ideal gas of photons, is another example. In this case the photon energy is inversely proportional to the wavelength. In addition, many real systems act as if their interparticle interactions are very weak except at extremely short distances, so that an ideal gas model is a good approximation to reality. This situation arises, for example, with electrons in metals. Finally, there are systems where the interactions are important, but some qualitative understanding of real systems can be obtained from the ideal gas model. An example of this situation is the superfluid transition of liquid helium. Clearly, since helium is a liquid at very low temperatures, the ideal gas model cannot be very accurate. Nevertheless, theory predicts the condensation of ideal bosons into a ground state that is analogous to the superfluid state of helium.

Much progress in thermodynamics and statistical mechanics has been obtained by working entirely with the so-called classical ideal gas. In this classical approximation not only is the energy totally kinetic, but also the effects of quantum theory are ignored. We will include the effects due to quantum theory, and the classical ideal gas will be the limiting case for high temperatures.

For our purposes the two main results of quantum theory are that the states of bound systems are discrete and identical particles are indistinguishable. This latter fact leads to the classification of all particles into two types: bosons and fermions. Fermions and bosons have very different behavior at low temperatures, which follows from the fact that two fermions cannot exist in the same single particle state, while bosons can. The statistics describing the two types of particles are called Fermi-Dirac (FD) for fermions and Bose-Einstein (BE) for bosons. At high temperatures all particles behave the same because the occupation of the same state by more than one particle is so rare that we can treat the particles as distinguishable. This approximation leads to a type of statistics called Maxwell-Boltzmann (MB). The occupation number or mean number of particles in any single particle state labeled by the index i, with energy ϵ_i, and chemical potential μ is given by

$$f_{\mathrm{FD}} = \frac{1}{e^{(\epsilon_i - \mu)/k_B T} + 1} \tag{6.1}$$

$$f_{\mathrm{BE}} = \frac{1}{e^{(\epsilon_i - \mu)/k_B T} - 1} \tag{6.2}$$

$$f_{\mathrm{MB}} = \frac{1}{e^{(\epsilon_i - \mu)/k_B T}}, \tag{6.3}$$

where k_B is Boltzmann's constant and μ is usually a slowly varying function of temperature. At very high temperatures, the chemical potential becomes very large in magnitude and negative, making the exponential in the denominator much larger than unity and thus both FD and BE statistics become equivalent to MB statistics. For bosons we can see that the chemical potential must be negative at all temperatures to insure that $f_{\mathrm{BE}} > 0$. A negative occupation number would not make sense physically. For fermions and distinguishable particles, the chemical potential can be positive or negative. Typically for electrons in metals the chemical potential does not vary much with temperature under laboratory conditions and is hundreds of times larger than $k_B T$. From Eq. 6.1, we see that at $T = 0$ the occupation number for states with energy below the chemical potential is unity and for states with energy above the chemical potential, it is zero. The state occupancy at room temperature does not change much from the $T = 0$ behavior except for energies close to the chemical potential.

Another factor that affects macroscopic properties is whether or not the particles have a mass (such as typical gases and electrons in metals) or are massless (such as photons or phonons). The difference between non-zero mass and massless particles manifests itself most directly in the chemical potential. For massless particles there is no conservation of the number of particles. If one considers the phonons in a crystal or the photons in a heated cavity, there is no reason why their number cannot fluctuate. For non-zero mass particles, however, we know that unless the particles obtain energies close to their rest energy, they cannot be destroyed or created. Because there is no energy barrier to changing the particle number for massless particles, their chemical potential must be zero.

Also, typically non-zero mass and massless particles differ in the relation between their energy and momentum, sometimes called the dispersion relation. For photons and long-wavelength phonons the energy is proportional to the momentum, while for non-relativistic particles, the energy is proportional to the square of momentum. The complete dispersion relation for phonons is more complicated and depends on the lattice structure of the crystal. (See the CUPS solid state programs.) Another effect of the crystal structure is that there is an upper limit to the energy, called the Debye energy, because the smallest distance and thus the smallest wavelength cannot be less than the lattice spacing. The crystal structure in metals also will affect the dispersion relation for the electrons. The periodicity of a crystal will impose a periodicity on the electron eigenfunctions and eigenvalues.

Finally, the dimension of space, d, is important, and particles restricted to a two-dimensional plane typically will have different behavior than those in three dimensions. Much interesting physics currently is being done in such restricted geometries. For example, some of the high-temperature superconductors behave as if they are stacks of independent two-dimensional planes.

Even though there is no potential energy of interaction between ideal gas particles, the calculation of the thermodynamic properties is still non-trivial and can necessitate numerical calculations. Thus, the computer becomes a useful tool. In addition, if the chemical potential is non-zero, it is very difficult to perform analytical calculations in the canonical ensemble where the temperature T, volume V, and the number of particles N is fixed. The reasons for the difficulties are discussed in section 6.3, and a computer simulation method using the canonical ensemble will be presented in section 6.4. Before we go through this calculation we summarize the results from quantum theory that we will need.

6.2 Review of Quantum Theory

We will assume that the particles are enclosed in a box of length L in d dimensions. The "volume" of such a box is $V = L^d$. In most textbook derivations, it is assumed that $d = 3$. However, we want to be more general, both because there are real systems that act as though they are in lower dimensions, and because we wish to see how dimensionality affects the results.

We wish to solve Schrödinger's equation in a box to determine the possible single particle states. To do so we must specify the boundary conditions. We could either assume that the wave function vanishes at the walls of the container or we could assume periodic boundary conditions. In one dimension this latter possibility amounts to replacing a line segment of length L with a circle of circumference L. The two-dimensional example would replace an $L \times L$ square by the surface of a torus. Both types of boundary conditions lead to the same macroscopic results. We will use periodic boundary conditions because it will turn out that the calculations are slightly simpler. Also for simulations periodic boundary conditions are usually preferable because the lack of a physical boundary means that all positions in space are equivalent.

The eigenfunctions for a non-relativistic particle of mass m in a box with periodic boundary conditions are[6]

$$\psi_{\mathbf{k}} = e^{i\mathbf{k}\cdot\mathbf{r}}, \tag{6.4}$$

where the wave vector $\mathbf{k}$ takes on the values

$$\mathbf{k} = \frac{2\pi}{L}(n_1\mathbf{e}_1 + n_2\mathbf{e}_2 + \ldots + n_d\mathbf{e}_d). \tag{6.5}$$

The energy eigenvalues are

$$\epsilon(\mathbf{k}) = \frac{\hbar^2 k^2}{2m} \qquad \text{non-zero mass.} \tag{6.6}$$

For massless particles such as photons and phonons at long wavelength,

$$\epsilon(\mathbf{k}) = \hbar k c \qquad \text{massless,} \tag{6.7}$$

where c is the particle speed. Here the n_i are positive or negative integers including 0, and the $\mathbf{e}_i$ are the usual orthogonal unit vectors. The values for $\mathbf{k}$ are determined by the periodic boundary condition:

$$\psi(\mathbf{r} + L\mathbf{e}_i) = \psi(\mathbf{r}). \tag{6.8}$$

What are the eigenfunctions for many particles? For noninteracting particles we can write the eigenfunctions for all the particles in terms of the eigenfunctions for individual particles such that the total energy eigenvalue is the sum of the single particle energy eigenvalues and the particles are indistinguishable. The condition of indistinquishability implies that if we denote any wave function for N particles by

$$\psi = \psi(\mathbf{k}_1, \mathbf{k}_2, \ldots, \mathbf{k}_N), \tag{6.9}$$

then

$$\begin{aligned} &|\psi(\mathbf{k}_1, \mathbf{k}_2, \ldots, \mathbf{k}_i \ldots, \mathbf{k}_j, \ldots, \mathbf{k}_N)|^2 \\ &= |\psi(\mathbf{k}_1, \mathbf{k}_2, \ldots, \mathbf{k}_j \ldots, \mathbf{k}_i, \ldots, \mathbf{k}_N)|^2. \end{aligned} \tag{6.10}$$

We label the wave function by the single particle wave vectors. The wave function on the right-hand side is identical to the one on the left, except that particle i has changed its state from $\mathbf{k}_i$ to $\mathbf{k}_j$ and particle j has changed its state from $\mathbf{k}_j$ to $\mathbf{k}_i$. Because only the square of the wave function, $\psi\psi^*$, has physical meaning, there cannot be any difference if we interchange any two particles. Thus, we have for the wave function itself

$$\psi(\mathbf{k}_1, \mathbf{k}_2, \ldots, \mathbf{k}_i, \ldots, \mathbf{k}_j, \mathbf{k}_N) = e^{i\phi}\psi(\mathbf{k}_1, \mathbf{k}_2, \ldots, \mathbf{k}_j, \ldots, \mathbf{k}_i, \ldots, \mathbf{k}_N), \tag{6.11}$$

where ϕ is an arbitrary phase angle. All particles in nature correspond to the case $e^{i\phi} = \pm 1$. The plus sign corresponds to bosons and the minus sign corresponds to fermions. If $\mathbf{k}_i = \mathbf{k}_j$ for fermions, we see that the wave function will equal minus itself and hence it must equal zero. This condition implies that two fermions cannot be in the same state, e.g., have the same value of $\mathbf{k}$. If we include the spin degrees of freedom, two fermions can have the same value of $\mathbf{k}$ if their spin values differ.

6.3 Grand Canonical Ensemble

6.3.1 Theoretical Background

We derive the basic formulae needed to calculate thermodynamic quantities using the grand canonical ensemble. First we will see why it is difficult to do the calculations in the canonical ensemble. To calculate macroscopic averages we need to compute the partition function

$$Z = \sum e^{-E(\mathbf{k}_1,\mathbf{k}_2,\ldots,\mathbf{k}_N)/k_BT}, \tag{6.12}$$

where E is the total energy, and the sum is over all possible states. We would like to write the sum as a product over all the single particle partition functions, but we cannot do so because the particles are indistinguishable. That is, it does not make sense to sum over the possible states of particle number 1, and then particle number 2, etc. Instead we can only sum over the single particle states and ask how many particles are in each state, such that the total number of particles equals N, and then sum over all the possible ways of distributing particles among states. This summation is difficult to do analytically.

However, in the grand canonical ensemble we are not restricted to a fixed number of particles. We can then assume each single particle state is like a separate system in equilibrium with all the other single particle states, that is, we associate the same chemical potential with each single particle state and determine how many particles are in each single particle state on the average. The number of particles in the single particle state labeled by $\mathbf{k}$ is

$$f_{\mathbf{k}} = \frac{\sum n\, e^{(\mu-\epsilon_{\mathbf{k}})/k_BT}}{\sum e^{(\mu-\epsilon_{\mathbf{k}})/k_BT}}, \tag{6.13}$$

where the sum is over $n = 0, 1$ for fermions and $n = 0, 1, 2, 3, \ldots$ for bosons. The sums can be done easily and they lead to Eqs. 6.1 and 6.2. We note that the number of particles in a single particle state depends only on its energy not explicitly on $\mathbf{k}$.

The next quantity we want to know is how many states have the same energy, since to find the thermodynamic quantities we will need to sum over the energy. The number of states in an energy interval ϵ to $\epsilon + d\epsilon$ per unit volume is called the density of states, $\rho(\epsilon)$. Let us derive the density of states by assuming that the energy is of the form (see Eqs. 6.6 and 6.7)

$$\epsilon = Bk^p. \tag{6.14}$$

We wish to find $\rho(\epsilon)$ for any spatial dimension d. To do so we imagine k-space to be divided up into cells of size $2\pi/L$, and ask how many states are within a hypersphere of radius k. The answer is

$$\Gamma(k) = n_s A_d (L/2\pi)^d k^d, \tag{6.15}$$

where n_s is the number of spin or polarization states a particle can have; A_d is a constant for a given dimension d and has the values

$$A_1 = 2 \tag{6.16}$$

$$A_2 = \pi \tag{6.17}$$

$$A_3 = 4\pi/3 . \tag{6.18}$$

Eventually, we will choose units so that n_s, A_d, and B drop out of our results. The density of states is given by the derivative of Eq. 6.15 divided by L^d:

$$\rho(k)\, dk = (1/L^d)\frac{d\Gamma}{dk}\, dk = n_s dA(1/2\pi)^d k^{d-1} dk \tag{6.19}$$

From Eq. 6.14 we have

$$d\epsilon = pBk^{p-1} dk \tag{6.20}$$

and

$$k = (\epsilon/B)^{1/p}, \tag{6.21}$$

and hence,

$$\rho(\epsilon)\, d\epsilon = \rho(k)\, dk = n_s A(d/p)(1/2\pi)^d (\epsilon/B)^{d/p-1} d(\epsilon/B) . \tag{6.22}$$

We would like to reduce this expression to one that is independent of constants that could always be added in at the end if needed. To do so we will define a special energy, ϵ_0, which is the Fermi energy for fermions, the Debye energy for phonons, but only sets the scale of energy for other systems. The energy ϵ_0 is defined by the relation

$$\rho \equiv \frac{N}{L^d} \equiv \int_0^{\epsilon_0} \rho(\epsilon)\, d\epsilon = n_s A(1/2\pi)^d\, (\epsilon_0/B)^{d/p}, \tag{6.23}$$

where N is the mean number of particles or for phonons is equal to the total number of energy levels, which is the number of degrees of freedom in the lattice. The number of degrees of freedom equals d times the number of atoms in the lattice, N_a. Hence for phonons we can interpret N as dN_a. If we use these results, we have

$$\epsilon_0 \equiv k_B T_0 = (\rho/n_s A)^{p/d} B(2\pi)^p . \tag{6.24}$$

Now if we measure all energies in terms of ϵ_0, then Eqs. 6.22 and 6.24 lead to

$$\rho(\epsilon)\, d\epsilon = \rho \frac{d}{p} \epsilon^{d/p-1} d\epsilon . \tag{6.25}$$

Now since the density of states is simply proportional to the number density, we can eliminate ρ by considering a density of states that is a number of states per particle (or for phonons per degree of freedom) rather than per volume. We will denote this density of states by $D(\epsilon)$. It is given by

$$D(\epsilon)\,d\epsilon \equiv (L^d/N)\rho(\epsilon)\,d\epsilon = \frac{d}{p}\epsilon^{d/p-1}d\epsilon\,. \tag{6.26}$$

This last result is the one we will use in the computer program. Remember that all energies are now in units of ϵ_0. Thus, if you see an energy of 1.5 on the computer screen, it corresponds to an energy in actual units of $1.5\epsilon_0$. Note that the temperature $T_0 \equiv \epsilon_0/k_B$ is the Fermi temperature for fermions and the Debye temperature for phonons. For all other systems T_0 is just a convenient reference temperature. In the following derivations we will always use the symbol T to represent the temperature using energy units. Thus the symbol T in energy units replaces k_BT with T measured in degrees Kelvin. With these units you will never see a $k_B, \hbar, n_s$, etc., on the screen. Temperatures shown on the screen are in units of T_0. For example, a temperature of 2.0 corresponds to $T/T_0 = 2.0$.

Once we have the density of states, $D(\epsilon)$ and the distribution function $f(\epsilon)$, it is easy to find many quantities of interest. The number of particles with an energy between ϵ and $\epsilon + d\epsilon$ per particle is

$$N(\epsilon)\,d\epsilon = f(\epsilon)D(\epsilon)\,d\epsilon\,. \tag{6.27}$$

The energy distribution of these particles is

$$I(\epsilon)\,d\epsilon = \epsilon f(\epsilon)D(\epsilon)\,d\epsilon\,. \tag{6.28}$$

For photons, $I(\epsilon)$ is proportional to the intensity of light from a blackbody. The total energy per particle is given by

$$E/N = \int_0^{\epsilon_{\max}} \epsilon f(\epsilon)D(\epsilon)\,d\epsilon, \tag{6.29}$$

where $\epsilon_{\max}$ is the Debye energy for phonons and is infinite otherwise.

The difficulty with Eqs. 6.27–6.29 for non-zero mass particles is that we do not know the chemical potential needed to compute f. The chemical potential is found by the implicit equation

$$N/V = \int_0^{\infty} f(\epsilon)\rho(\epsilon)\,d\epsilon \tag{6.30}$$

and thus from Eq. 6.26

$$1 = \int_0^{\infty} f(\epsilon)D(\epsilon)\,d\epsilon\,. \tag{6.31}$$

To solve this equation for the chemical potential we first note that the energy ϵ and the chemical potential both appear divided by T in the distribution function f. Let us define

$$y = \mu/T \tag{6.32}$$

and

$$x = \epsilon/T. \tag{6.33}$$

If we use this change of variables in our formulae for $D(\epsilon)$ and $f(\epsilon)$, and recall that all temperatures and energies (including μ) are to be measured relative to ϵ_0, we have

$$T^{-d/p} = \frac{d}{p}\int_0^\infty \frac{x^{d/p-1}\,dx}{z^{-1}e^x + s}. \tag{6.34}$$

Here the fugacity z is defined by

$$z = e^y, \tag{6.35}$$

and the integer s has the values

$$s = +1\ (FD) \tag{6.36}$$

$$= -1\ (BE) \tag{6.37}$$

$$= 0\ (MB). \tag{6.38}$$

Thus, given a y or z, we can use Eq. 6.34 to find the temperature. With y known, we can go back and do another integral to find the energy. If we rewrite the energy in terms of an integral over x, we obtain

$$\frac{E}{N} = \frac{d}{p}T^{d/p+1}\int_0^{x_{\max}} \frac{x^{d/p}\,dx}{z^{-1}e^x + s}. \tag{6.39}$$

The upper limit of integration in the Debye model is T_0/T; otherwise it is infinite. Notice that we have to compute an integral of the form

$$I(z, s, a, x_{\max}) = \int_0^{x_{\max}} \frac{x^a\,dx}{z^{-1}e^x + s} \tag{6.40}$$

for both the energy and chemical potential. In the Appendix (section 6.7) we discuss the various methods for numerically computing this integral.

The specific heat per particle is given by

$$c = \frac{1}{N}\frac{dE}{dT} = \frac{d}{p}(\frac{d}{p} + 1)\,T^{d/p}\,I(z, s, d/p, x_{\max}) \tag{6.41}$$

$$+\frac{d}{p}T^{\frac{d}{p}+1}\frac{dI(z, s, d/p, x_{\max})}{dT}.$$

For non-zero mass particles the temperature dependence of the last term comes from the chemical potential, whereas for phonons, it comes from the upper limit of integration. The last term in Eq. 6.41 is zero for photons for which $x_{\max} = \infty$, and for bosons below the Bose condensation temperature since $\mu = 0$.

Now we consider non-zero mass particles. By the chain rule, the derivative of I with respect to T can be obtained from the derivative of I with respect to z:

$$\frac{dI(z,s,d/p,\infty)}{dT} = I'(z,s,d/p,\infty)(dz/dT), \tag{6.42}$$

where the prime indicates a derivative with respect to z. The derivative dz/dT can be obtained by implicit differentiation of Eq. 6.34:

$$\frac{dz}{dT} = \frac{-T^{-d/p-1}}{I'(z,s,\frac{d}{p}-1,\infty)}. \tag{6.43}$$

The last term in Eq. 6.41 then becomes

$$-\frac{d}{p}\frac{I'(z,s,d/p)}{I'(z,s,\frac{d}{p}-1)}. \tag{6.44}$$

For phonons the last term in Eq. 6.41 is obtained by replacing the integral in Eq. 6.39 by its integrand evaluated at T_0/T. Thus this contribution to the specific heat becomes

$$-\frac{d}{p}\frac{(T_0)^{1+d/p}}{T}\frac{1}{e^{T_0/T}-1}. \tag{6.45}$$

One of the interesting effects we would like to find is Bose condensation. Bose condensation means that a macroscopic number of the particles are in the ground state. How do our calculations predict this effect? For bosons the limiting value of z is unity because the chemical potential must be negative. If we compute the temperature for $z = 1$, we may obtain a non-zero value. Because the temperature decreases monotonically with z, this means that we have reached a limiting temperature. This temperature, T_c, is the transition between a normal fluid of bosons and a condensed Bose fluid. Below this temperature $z = 1$ (i.e. the chemical potential is zero). In one and two dimensions with $p = 2$, we find that $T_c = 0$. In general,

$$T_c = \left(\frac{p/d}{I(1,-1,\frac{d}{p}-1,\infty)}\right)^{p/d}. \tag{6.46}$$

6.3.2 Procedure for Running Program QMGAS1

The goal of this program is to allow the user to generate thermodynamic data and the functions $D(\epsilon)$, $f(\epsilon)$, $N(\epsilon)$, and $I(\epsilon)$ at various temperatures for a variety of ideal systems. A program is needed to calculate the integrals $I(z,s,a,x_{\max})$, and to help organize the calculations. The steps in running the program are as follows:

1. Choose **System** from the menu. Then define the system from an input screen. (Alternatively, if you have previously generated a table of data, you can read that table in by selecting **Cancel** on the input screen and choosing **File/Open**.) The system is defined by the type of statistics (BE, FD, or MB), the relation between energy and wave number given by the exponent p, the dimension of space, whether the chemical potential is zero, and whether there is a Debye cutoff.

or reject a new state is identical to that used for the Ising model simulation in chapter 7.

As the system evolves from one state or configuration to another, we can calculate various average quantities over the configurations of particles. In our case a configuration is simply a list of how many particles are in each single particle state. The averages in the simulation should approximately equal the exact average over all possible states. Quantities we can calculate include the distribution function $f(E)$, the mean energy per particle E/N, and the specific heat c calculated from the fluctuations in the energy,[1-3] $c = (1/NT^2)(\langle E^2\rangle - \langle E\rangle^2)$, with units such that $k_B \equiv 1$. We use units such that the energy of a state with wave number k is given by $E = k^p$, where p is specified by the user, and the wave vector k has coordinates that are integers. These units correspond to setting $B = 1$ and $L = 2\pi$ in Eqs. 6.14 and 6.15. In these units all temperatures represent $k_B T/\epsilon_1$, where ϵ_1 is the first excited single particle energy level. We define a state $\mathbf{k} = 0$ to be the ground state with energy 0.

The state space or list of possible single particle states is defined by specifying the maximum value for each component of $\mathbf{k}$, and consists of a line segment in $d = 1$, an ellipse in $d = 2$, and an ellipsoid in $d = 3$. How do we change the configuration and not violate the principle of indistinguishability for quantum particles and in particular the Pauli exclusion principle for fermions? We have implemented the indistinguishability criteria by ordering all the single particle states from lowest energy to highest energy. States with equal energies are placed in an arbitrary order with respect to each other. Then a configuration is changed by moving a particle from one state to another such that the particle does not jump over an occupied state in the ordered list. In this way the particles are always ordered in the same sequence in the list throughout the simulation, and we are sampling only one ordering of the particles. If we had allowed a move that swapped two particles, then in actuality nothing would have changed. Thus, in program QMGAS2 we sample over the occupation numbers of the states rather than over the possible states for each particle. An illustration of a possible move is shown below in Figure 6.4.

The exclusion principle is implemented by beginning with no more than one particle in a single particle state (we ignore spin in this simulation), and we do not allow any move where any state is occupied by more than one fermion. For bosons we allow more than one particle in a state, but we do not allow the ordering of the particles in the list to change. To implement Maxwell-Boltzmann statistics, we allow the ordering of the particles to change, and hence they are distinguishable.

To simulate massless particles where the number of particles is not fixed, we can think of the zero energy state as a source and sink of particles, and only count those particles in excited states as actual particles. As long as the number of particles in the ground state remains large during the simulation, this procedure should be a reasonable approximation.

6.4.2 Procedure for Running Program QMGAS2

The simulation proceeds by choosing a particle at random and attempting to move it according to the Metropolis algorithm. One Monte Carlo step (MCS) is defined

2. Obtain data. If the chemical potential is set to zero, you will see an input screen asking for a temperature. Otherwise you will have to generate a temperature from the quantity $y = \mu/T$ for FD and MB statistics, or from $z = e^{\mu/T}$ for BE statistics. If the Bose condensation temperature is non-zero, you also can choose a reduced temperature $t = T/T_c \leq 1$. After each temperature is chosen or calculated, the data for μ, E/N, and c is inserted into a list ordered by increasing temperature, and the updated list is shown. Choose **Cancel** from the input screen when you have enough data. You may add more data later. A sample list of data is shown in Figure 6.1. After you have chosen at least two points you can do any of the following by choosing the appropriate hot key:

3. a. Add more data points to the list by choosing **F5-Add Data**.
   b. Delete data points from the list by choosing **F6-Delete**.
   c. Look at $D(\epsilon)$, $f(\epsilon)$, $N(\epsilon)$, and $I(\epsilon)$ for three different temperatures by choosing **F2-E Plots**. A sample screen showing the distribution functions is given in Figure 6.2.
   d. Look at plots of μ, E/N, and c versus temperature by choosing **F3-T Plots**. The scales on the axes are produced automatically by the CUPS routines, but usually they are not very convenient. After the plots appear, use **F3-Rescale** to set your own scales. A sample set of plots produced by program QMGAS1 is shown in Figure 6.3.

4. When you wish to change the system parameters, choose **System** from the menu.

5. When you wish to exit the program, choose **File** from the menu and select **Exit Program**.

In Figure 6.1 we show the list of data for a BE system with $d = 3$, $p = 2$, non-zero chemical potential, and no Debye cutoff. Eleven temperatures are shown. Figure 6.2 shows the plots for $D(\epsilon)$, $f(\epsilon)$, $N(\epsilon)$, and $I(\epsilon)$ for three temperatures drawn from the original list. In the program each temperature would be drawn in a separate color. Note that $D(\epsilon)$ is the same for all temperatures. Finally, Figure 6.3 shows the plots of chemical potential, the energy per particle, and the specific heat for the data in the list of Figure 6.1.

6.4 Canonical Ensemble Simulation

6.4.1 Monte Carlo Algorithm

Calculations in the canonical ensemble are difficult to do because we must be able to compute sums or integrals with the constraint that the number of particles is fixed. Numerical procedures do exist for noninteracting systems with a small number of particles to calculate the thermodynamic properties exactly. Because quantum particles are indistinguishable, these methods involve being able to compute the number of ways of permuting or exchanging all possible combinations of particles.

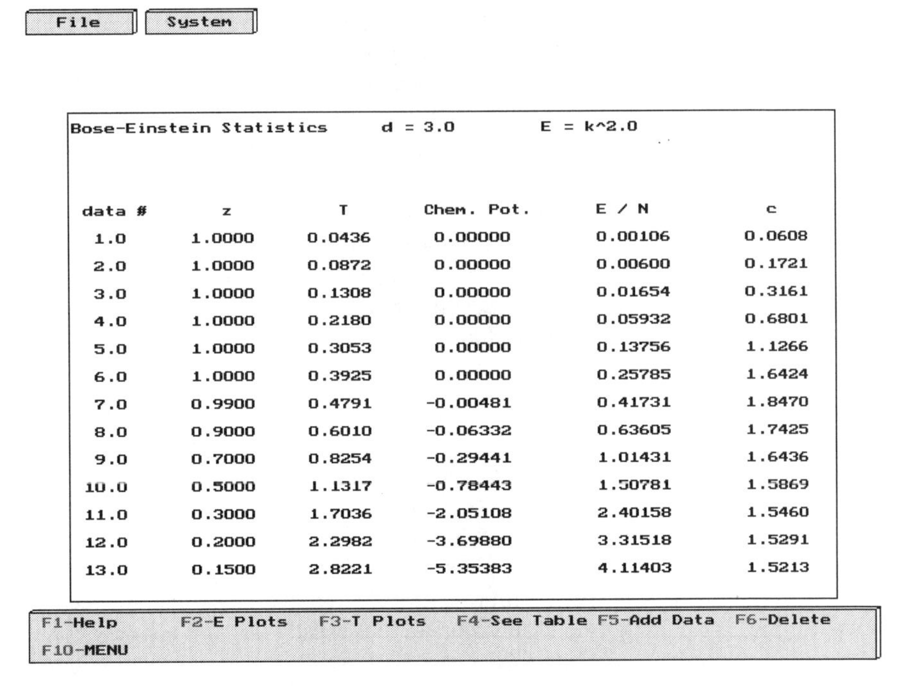

File | System

Bose-Einstein Statistics d = 3.0 E = k^2.0

data #	z	T	Chem. Pot.	E / N	c
1.0	1.0000	0.0436	0.00000	0.00106	0.0608
2.0	1.0000	0.0872	0.00000	0.00600	0.1721
3.0	1.0000	0.1308	0.00000	0.01654	0.3161
4.0	1.0000	0.2180	0.00000	0.05932	0.6801
5.0	1.0000	0.3053	0.00000	0.13756	1.1266
6.0	1.0000	0.3925	0.00000	0.25785	1.6424
7.0	0.9900	0.4791	-0.00481	0.41731	1.8470
8.0	0.9000	0.6010	-0.06332	0.63605	1.7425
9.0	0.7000	0.8254	-0.29441	1.01431	1.6436
10.0	0.5000	1.1317	-0.78443	1.50781	1.5869
11.0	0.3000	1.7036	-2.05108	2.40158	1.5460
12.0	0.2000	2.2982	-3.69880	3.31518	1.5291
13.0	0.1500	2.8221	-5.35383	4.11403	1.5213

F1-Help F2-E Plots F3-T Plots F4-See Table F5-Add Data F6-Delete
F10-MENU

Figure 6.1: Typical list of data produced for a system of bosons in three dimensions.

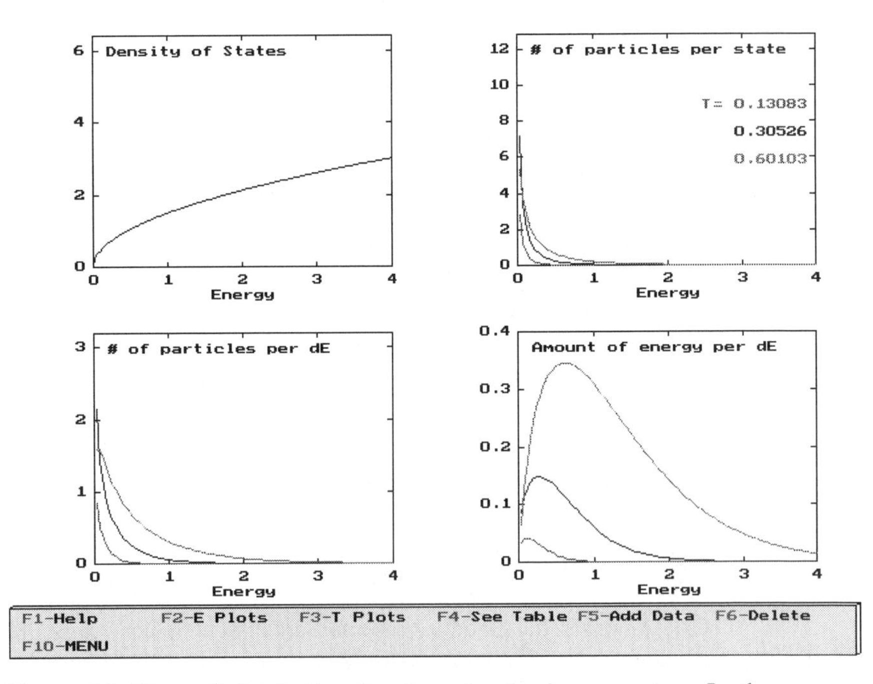

Figure 6.2: Plots of distribution functions for the boson system. In the program curves for each temperature have a different color.

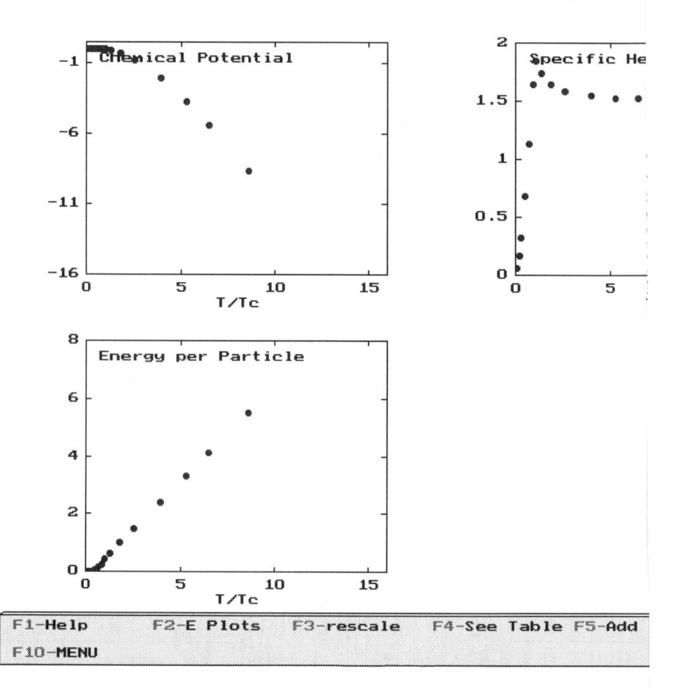

Figure 6.3: Plots of thermodynamic data for boson system. Tl have been set by using the **rescale** hot key.

Such methods do not provide a very intuitive look at what tl doing. The procedure we will use, known as the Metropolis M is not very efficient computationally for the quantum ideal g some insight into the system at the microscopic level.

The simulation will be done in momentum space. Inste particles to be moving in a box, we will imagine they are mo particle state to another, where each state is specified by the

Typically Monte Carlo simulations of thermodynamic syste to evolve from one state to another so that the distribution c equilibration) is given by the correct Boltzmann weight. On initial state, and then uses an algorithm to change the state. To c distribution, it is sufficient that the transition probability, $P(i$ to state j is related to the reverse probability by

$$P(i \longrightarrow j) = P(j \longrightarrow i)e^{-\beta(E_i - E_j)},$$

where $\beta = 1/k_B T$, and E_i is the energy of the ith state. Thi *detailed balance*. The Metropolis algorithm for evolving the sys detailed balance proceeds as follows. First, choose a new state change in energy, $\Delta E = E_{\text{new}} - E_{\text{old}}$. If $\Delta E \leq 0$, accept the r accept it with probability $P = e^{-\beta \Delta E}$. This last condition is computer by comparing P with a random number, r, uniformly 0 and 1 and accepting the change if $P > r$. One can show that the requirement of detailed balance. This procedure for decidin

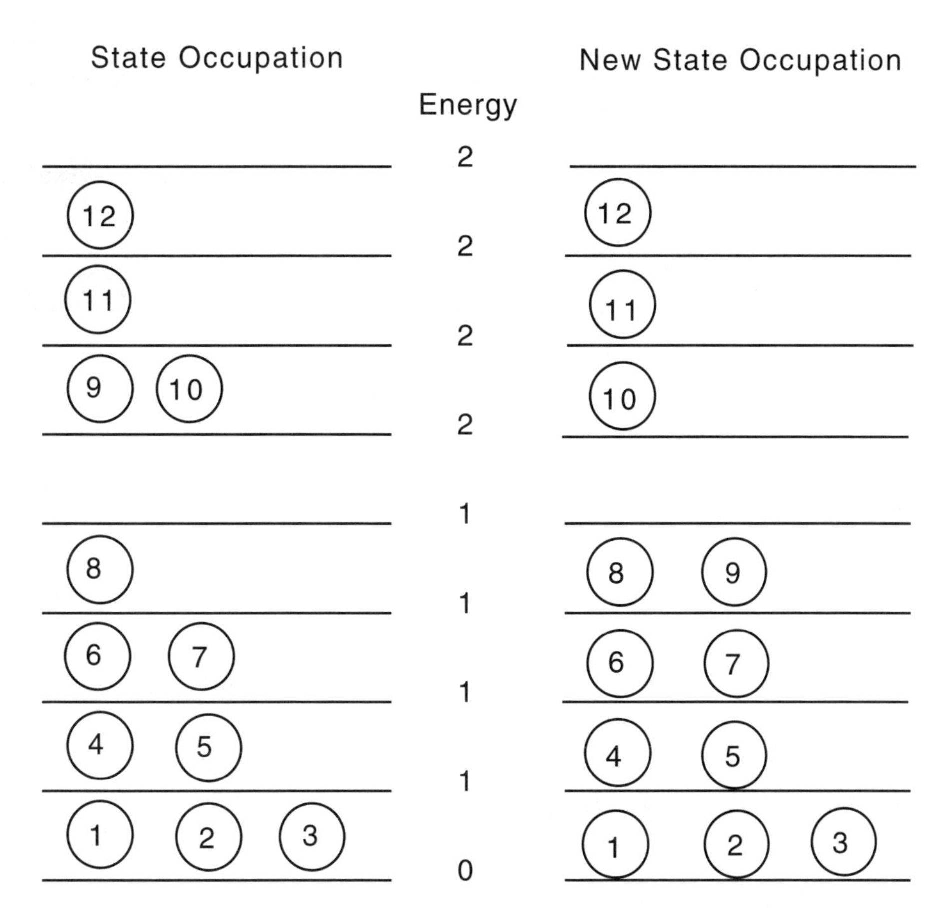

Figure 6.4: Illustration of a possible Monte Carlo move for bosons. Each horizontal line represents an energy state. The column of numbers at the center gives the energy of each state. Note that most energy levels are degenerate, meaning that there is more than one energy state with the same energy. Each boson is assigned a number. Note that boson #I always has at least as much energy as boson #J if $I > J$. Thus, the ordering of the particles never changes throughout the simulation. The Monte Carlo move consists of particle #9 moving to a lower energy state.

as N attempted moves, where N is the number of particles. After each MCS, data is collected for the averages. The stages in running this program are as follows:

1. An initial input screen will appear from which you may choose the type of statistics (BE, FD, or MB), size of momentum space in each direction, number of particles N, exponent relating energy to wave number, p, and the initial temperature. The simulation then shows a representation of momentum space where the occupation of each state is indicated by a color for different occupation numbers. In one and two dimensions the entire momentum space is shown. In three dimensions the $k_z = 0$ plane is shown. Also shown is the instantaneous and average energy per particle versus the number of MC steps replotted every 1000 MCS.

2. After choosing the system parameters, hit the hot key **F6-RUN** to start the simulation. At any time during the simulation you can hit the hot key **F6-PAUSE** to stop the simulation.

3. While the simulation proceeds, you can click on **F2-RESET** to reset the accumulation of data for the averages.

4. Choose **F3-UPDATE** to see the current thermodynamic averages, the acceptance rate, and a plot of $f(\epsilon)$. Make sure you choose **F3-UPDATE** only after the simulation has started.

5. Change the temperature with the slider. Whenever the temperature is changed, the accumulation of data is reset.

6. Choose **F4-SLOWER** to slow down the simulation and **F5-FASTER** to speed up the simulation.

7. Choose **System** from the menu to change the system or to set the temperature higher than the maximum on the slider.

8. Choose **File** from the menu and then click on **Exit Program** to exit from the program.

6.5 Exercises

Grand Canonical Ensemble

6.1 **Wien's Law**
Use the program to verify the Wien displacement law for photons. According to this law the maximum in the $I(\epsilon)$ curve is proportional to T. Find the proportionality constant. Is Wien's law valid in any dimension of space? Test it for $d = 1$ and 2. If the energy was proportional to the square of the wave number k, would Wien's displacement law still be true? If so, is the proportionality constant the same?

6.2 **Bose-Einstein Condensation Temperature**
Find T_c for non-zero mass bosons in $d = 1, 1.5, 2, 2.5, 3, 3.5, 4$ dimensions. Fractional dimensions can be realized by embedding a substance in a porous medium that has a geometry such that volumes scale as if there were a fractional dimension. Such structures are called fractals, and experiments on helium in the porous medium called vycor have recently been done. Give a qualitative reason for the dependence you find. What happens if you change the energy relation to $E \propto k$?

6.3 **Fermion Specific Heat**
Estimate graphically from the $f(\epsilon)$ plot the specific heat for a Fermi system at low temperatures by estimating how much energy has moved from below the Fermi level to above the Fermi level for different temperatures. Compare your estimate to that calculated directly by the program.

6.4 **Fermion Distributions**

Use the program to estimate at what temperature, compared to the Fermi temperature, more than 10% of the particles are above the Fermi energy. Does this temperature depend on the dimension d or the parameter p, where $E \propto k^p$? Does the integral under the $f(\epsilon)$ curve always equal unity in our units or is this behavior true only at low temperatures? Explain your answer. What happens in one dimension (with $p = 2$)?

6.5 **Classical Limit**

Compare the distribution functions for BE statistics with those for MB statistics at various temperatures and estimate the temperature at which quantum effects become unimportant. Repeat for FD statistics. Is it the same for bosons and fermions? Do the same estimate from the specific heat data. What should the specific heat be at high temperatures?

6.6 **Debye Crystal**

Determine the temperature at which the low temperature T^3 law for the specific heat sets in for phonons in three dimensions. At what temperature do phonons begin to behave as a classical solid where the specific heat as a function of temperature is a constant? Suppose we crudely mimic a more realistic dispersion relation by setting $\epsilon \propto k^{1.3}$. How does this relation affect the results? What is the low temperature form for the specific heat? Is the specific heat constant at high temperatures? If so, does this constant depend on how ϵ depends on k?

6.7 **Massless Particles**

Use the program to compare the physics of phonons (which have the Debye cutoff) and photons. Does the cutoff affect the high temperature or the low temperature properties? Explain your result in terms of the difference between phonons and photons for the energy of the largest k states. How do the distribution functions compare? How does the density of states compare?

6.8 **Chemical Potential**

Use Program QMGAS1 to calculate the chemical potential as a function of temperature for BE, FD, and MB gases. Explain the sign of the chemical potential and their relative magnitudes in the three cases in terms of the properties of the three kinds of particles. Recall that the chemical potential is a measure of the difficulty in adding a particle. Repeat this exercise for the comparison of the temperature dependence of the chemical potential in one, two, and three dimensions.

Canonical Ensemble Simulation

6.9 **Comparing Statistics**

Run a FD, BE, and MB system at $T = 1.5$ in $d = 2$. Describe qualitatively the occupancy in k-space, and discuss how it is consistent with what we expect from the distribution functions, Eqs. 6.1 to 6.3. Change the temperature and describe how the k-space occupancy changes for each kind of statistics. For which system does the occupancy of various states change the least(most) as the temperature is raised?

6.10 **Bose Condensation**

For finite systems there are no true phase transitions. However by comparing results for different sized systems, we usually can see the effects of a phase transition. For a three-dimensional system with $k_{\max} = 4$ in each direction, determine the ground state occupancy, $f(0)$, as a function of the number of bosons for $T = 1.0$. Try $N = 10, 20, 40$, and 80. Because $\epsilon = 0$ for the ground state, we can estimate the chemical potential from $f(0)$. For a thermodynamic system with many particles $\mu \propto -T/N$ below T_c. Do the simulation to see if this relation is applicable for small systems. If not, explain what could cause the difference. Try other temperatures and other sizes. In $d = 1$ and $d = 2$, we do not expect any phase transition. Is the dependence of μ at low temperatures different in these lower dimensions?

6.11 **Chemical Potential for Fermions**

When $\epsilon = \mu$ the distribution function $f(\epsilon) = 0.5$. Use this fact and the simulation program to find μ and to estimate its temperature dependence in $d = 2$ for 50 particles with $k_{\max} = 8$ in each direction. At what temperature does μ change its value by 20% from its low temperature value? What is the Fermi temperature for this system?

6.12 **Fermions in Different Dimensions**

The Fermi temperature is the greatest energy of an occupied state at zero temperature. Set the Fermi energy at $\epsilon_F = 9$ and determine the number of particles needed to obtain this energy for $d = 1$, 2, and 3. Use these particle numbers and $k_{\max} = 6$ in each direction for the simulations. Run the simulation and record the equilibrium specific heat at three widely different temperatures in each dimension. How do the results depend on dimensionality? Explain your results qualitatively using the k-space occupancy plots.

6.13 **Quenched Quantum Systems**

Run a $d = 2$ FD simulation of 50 particles in a system with $k_{\max} = 8$ in each direction, and set the temperature to $T = 100$. After equilibrium has been reached so that the energy fluctuates about its mean value, quench the system immediately to a temperature $T < 5$. What happens? Estimate the time for the system to reach its low temperature equilibrium. State your criteria for equilibrium. Repeat for different starting temperatures. Repeat for a BE and then for MB statistics. Compare your results.

6.14 **Effect of Particle Mass**

Use the Monte Carlo program to compare the effect of changing the energy relation from $\epsilon = k^2$ appropriate to non-zero mass particles, and $\epsilon = k$ for massless particles. Use BE statistics in 2D, $k_{\max} = 8$, and a large number of particles (try 200) so that the ground state is never depleted (this condition should simulate zero chemical potential for massless particles). How does the energy and the specific heat versus temperature compare in the two cases? How do the distribution functions differ?

6.15 **Effect of k-Space Shape**

Use the Monte Carlo program to compare the effect of allowing $k_{\max}$ in the x direction to be different from its value in the y direction. Begin with

BE statistics in 2D, $k_{x,\max} = 10$, and $k_{y,\max} = 2$. Is there any difference between this case and the symmetrical case? Repeat this simulation with FD statistics. For what physical situation would this calculation be relevant?

6.16 **Effect of Debye Cutoff**
We can model a Debye cutoff by using a k-space small enough that particles reach the edge of k-space. Use the program to compute the specific heat for a 100 particle BE system in $d = 2$ with $\epsilon = k$ for a number of temperatures. Compare the results for $k_{\max} = 8$ with those for $k_{\max} = 2$. Does the cutoff affect high or low temperature properties? Based on what you see with the k-space occupancy plot, explain your answer.

6.6 Program Modifications

The most important modification for both parts of the program is to allow for more realistic dispersion relations. For example, for a simple cubic lattice with lattice constant a and only harmonic interactions, the dispersion relation for phonons is

$$\epsilon(\mathbf{k}) = \hbar \frac{c_s}{a} \sqrt{2(1 - \cos ka)}, \tag{6.48}$$

where c_s is the sound speed in the long wavelength limit. For small k this result leads to the form $\epsilon \propto k$ used in the program. For particles moving with relativistic speeds and momentum $p = \hbar k$, we have

$$\epsilon(\mathbf{k}) = \sqrt{p^2c^2 + m_0^2c^4}. \tag{6.49}$$

In the non-relativistic limit the first term in the square root is much smaller than the second, and the result leads to ϵ equal to the rest energy plus $p^2/2m_0$. Since the rest energy is the same for all energy levels in this limit, we recover, by a shift in the zero of energy, the form $\epsilon \propto k^2$ used in the program.

It is easy to make these modifications in the Monte Carlo simulation program by just changing the energy computation in Procedure **SortLevels**. For the grand canonical calculation the function entering Simpson's rule to calculate the integrals and the special cases (Eqs. 6.51 to 6.55) would need to be modified. Also the simplified units used to avoid the various fundamental constants would need to be modified.

6.7 Appendix

Here we describe the numerical approximations used to evaluate the integral:

$$I(z, p, a, x_{\max}) = \int_0^{x_{\max}} \frac{x^a\, dx}{z^{-1}e^x + s}, \tag{6.50}$$

where $z \equiv e^y$. In a number of special cases there exist approximations that are better than standard integration algorithms. Except for those cases, we use Simpson's rule to do the integral.

If $z < 1$ or $y < 0$ and $x_{\max} = \infty$, we can multiply the numerator and denominator by z and then expand the integrand in a power series each term of which can be integrated. In this way we would find the series[3]

$$I(z, s, a, \infty) = \Gamma(a + 1) \sum_{m=1}^{\infty} \frac{(-s)^{m+1} z^m}{m^{a+1}}, \tag{6.51}$$

where $\Gamma(x)$ is the Gamma function that for non-negative integer values of x is given by $\Gamma(x + 1) = x!$ For very large values of z with $s = 1$(Fermions), there is a special expansion known as the Sommerfeld expansion given by[5]

$$I(z = e^y, s, a, \infty) = \frac{y^{a+1}}{a + 1} \sum_{n=1}^{\infty} a(a - 1) \cdots (a - 2n + 2) y^{a+1-2n} \sum_{m=1}^{\infty} \frac{2(-1)^{m+1}}{m^{2n}}. \tag{6.52}$$

In some cases the specific heat involves two terms of similar magnitude such that a straightforward use of the Sommerfeld expansion leads to incorrect results for very large values of y. For $y > 5$, we use a direct approximation for Eq. 6.41 given by the first couple of terms:

$$c = \frac{b\pi^2}{3y} + (2.705808(-4b^3 + 6b^2 + 2) + 1.894065616(12b^3 - 36b^2 + 24b))/y^3, \tag{6.53}$$

where $b = d/p$. There are a few analytical cases. For MB statistics, the integral is given by

$$I(z, 0, a, \infty) = \Gamma(a + 1)z. \tag{6.54}$$

If $a = 0$ then for all statistics

$$I(z, s, 0, \infty) = s \ln(1 + sz). \tag{6.55}$$

References

1. Reif, F. *Fundamentals of Statistical and Thermal Physics,* New York: McGraw-Hill, 1965.

2. Kittel, C., Kroemer, H. *Thermal Physics.* 2nd ed. San Francisco: W. H. Freeman, 1980.

3. Callen, H. B. *Thermodynamics and an Introduction to Thermostatistics.* New York: John Wiley and Sons, 1985.

4. Gould, H., Tobochnik, J. *An Introduction to Computer Simulation Methods, Part 2.* Reading, MA: Addison-Wesley, 1988.

5. Ashcroft, N. W., Mermin, N. D. *Solid State Physics.* Holt, Rinehart, and Winston, 1976.

6. Liboff, R. L. *Quantum Mechanics.* Reading, MA: Addison-Wesley, 1992.

7

The Ising Model and Critical Phenomena

Harvey Gould

7.1 Introduction

One of the remarkable properties of macroscopic systems is the existence of phase transitions. For example, a system of argon atoms can exist as a gas, liquid, or crystalline state, but in each state the microscopic interaction between pairs of atoms is exactly the same. How does the interaction between individual atoms lead to different phases?

In chapter 5 we studied some of the different phases of a simple model of argon. In principle, we also could use a computer to study the same model near a phase transition, e.g., near the liquid-gas critical point or near the transition from a liquid to a solid. However, such a study would require systems much larger than can be simulated in a relatively short time using today's personal computers. Instead, we will study phase transitions in a simple model of a magnetic system. In this system there is only one transition, that is, the system goes from a paramagnet to a ferromagnet at a well-defined temperature in zero external magnetic field. Although we will describe the phase transition in terms of magnetic language, the nature of the gas-liquid transition is very similar to the paramagnetic to ferromagnetic transition.

The model of magnetism that we will consider, the Ising model, is one of the most studied models in physics. The model is not discussed in detail in most undergraduate textbooks, but is discussed in references **?** and **?**, for example. One reason for its recent popularity is that it can easily be simulated on a computer. We introduce the Ising model in section 7.2 and discuss some of the theoretical background in sections 7.2–7.6.

7.2 The Ising model

Consider a solid of N identical atoms arranged on a regular lattice, and suppose that each atom has a net electronic spin $\mathbf{S}$ and an associated intrinsic magnetic moment μ. If we apply an external magnetic field $\mathbf{B}$, the energy of interaction with the magnetic field is $-g\mu_0 \sum_{i=1}^{N} \mathbf{S}_i \cdot \mathbf{B}$, where μ_0 is the Bohr magneton and $\mathbf{S}_i$ is the spin of the atom at lattice site i. In general, the dominant interaction between atoms is the "exchange" interaction, a consequence of the Pauli exclusion principle. The form of the exchange interaction for atoms i and j is $-J\mathbf{S}_i \cdot \mathbf{S}_j$, where J measures the strength of the exchange interaction. If $J > 0$, it is energetically favorable for the two spins to be parallel. Similarly, if $J < 0$, it is energetically favorable for the two spins to be antiparallel. The exchange interaction depends on the relative positions of the atoms, but is large only when their separation is one or two lattice spacings.

Because the spin and angular momentum are quantum mechanical operators, a theoretical treatment of magnetism involves all the complications of quantum mechanics. We will make several approximations that preserve the essential physics of the phase transition, but allow us to treat the system classically. The main approximation is to consider only the component of the spins along the z direction, the direction of the magnetic field. For simplicity, we also assume that the exchange interaction acts only between nearest neighbors. The energy of interaction for a system of N atoms can then be written in the form (cf. ref. **?**)

$$E = -J \sum_{i<j} s_i s_j - H \sum_i s_i \,, \tag{7.1}$$

where the sum in the first term is only over nearest neighbors. To simplify the notation, we have adopted the notation $s_i = \pm 1$, where s_i is the z-component of the spin (spin quantum number) at lattice site i. We say that $s_i = +1$ is an "up" spin and $s_i = -1$ is a "down" spin. All factors of μ_0 and g are incorporated into H and J (which both have dimensions of energy). The form of the energy of interaction in Eq. 7.1 is known as the Ising Hamiltonian.

The Ising model is not an exact description of magnetism, e.g., it is not applicable at low temperatures, but it is a non-trivial model of cooperative phenomena near a phase transition. The interaction between the spins makes a theoretical treatment very complicated, despite its simple form. The thermodynamic properties of the Ising model can be solved exactly in one dimension as a function of T and H and in two dimensions when $H = 0$. The theoretical treatment in two dimensions is a mathematical tour de force, but unfortunately does not provide much physical insight. The properties of the system in three dimensions have not been solved exactly. Fortunately, computer simulations and approximate theoretical treatments provide much physical insight.

7.3 One-Dimensional Ising Model

The thermodynamic properties of the Ising model can be calculated most easily in the canonical ensemble, that is, for fixed T, N, and H. Our goal is to enumerate all the possible microstates (denoted by s) and their corresponding energy E_s, calculate the partition function $Z_N = \sum_s \exp(-\beta E_s)$ ($\beta = 1/kT$), and then take the limit $N \longrightarrow \infty$. The problem is that the number of microstates, 2^N, becomes too large to enumerate for $N >> 1$. However, for the one-dimensional Ising model, we can calculate Z_N analytically for finite N and then take the limit $N \longrightarrow \infty$.

For $H = 0$, it is convenient to choose free boundary conditions so that the spins at each end have only one interaction. The energy can be written as

$$H = -J \sum_{i=1}^{N-1} s_i s_{i+1} \,. \tag{7.2}$$

We begin by considering two spins. There are four possible states: both spins up, both spins down each with energy $-J$, and two states with one spin up and one spin down each with energy $+J$. Thus Z_2 is given by

$$Z_2 = 2e^{\beta J} + 2e^{-\beta J} = 4\cosh \beta J \,. \tag{7.3}$$

In the same way we can enumerate the eight microstates for $N = 3$ and find that (see Exercise 7.1)

$$Z_3 = (e^{\beta J} + e^{-\beta J})Z_2 = 2Z_2 \cosh \beta J \,. \tag{7.4}$$

These considerations might convince you that the general relation between Z_N and Z_{N-1} is

$$Z_N = 2Z_{N-1} \cosh \beta J = 2(2\cosh \beta J)^{N-1} \,. \tag{7.5}$$

The relation 7.5 is derived in Exercise 7.2.

We can use the general result Eq. 7.5 for Z_N to find the Helmholtz free energy:

$$F = -kT \ln Z_N = -kT[\ln 2 + (N - 1)\ln(2\cosh \beta J)] \,. \tag{7.6}$$

In the thermodynamic limit $N \longrightarrow \infty$, the term proportional to N in Eq. 7.6 dominates, and we have

$$F = -NkT \ln(2\cosh \beta J) \,. \tag{7.7}$$

If we had calculated the thermodynamic properties in a non-zero magnetic field, we would have found that m, the magnetization per spin, is unity (all spins up) for $H \geq 0$ at $T = 0$, but $m = 0$ for $T > 0$ for $H = 0$. Hence there is no phase transition for the one-dimensional ($d = 1$) Ising model except at $T = 0$. Exercise 7.3 is devoted to the exploration of the thermodynamic properties of the $d = 1$ Ising model.

7.4 Mean Field Theory

Mean field theories predict phase transitions and some of their characteristic features. However, these theories are quantitatively incorrect, which is not surprising given the approximations that are made. In particular, we will find that mean field theory is insensitive to the spatial dimension and in its simplest form, incorrectly predicts a phase transition in one dimension.

The simplest form of mean field theory assumes that each spin interacts with the same effective magnetic field due to the external magnetic field and the internal field due to all the other spins. The effective field H_i at spin i is given by

$$H_i = -J \sum_{j=nn(i)} s_j - H, \tag{7.8}$$

where the sum in Eq. 7.8 is over the q nearest neighbors of i. Because the orientation of the neighboring spins depends on the orientation of spin i, H_i fluctuates from its average value, which is given by

$$\langle H_i \rangle = -H - J \sum_{j=1}^{q} \langle s_j \rangle = -H - qJm. \tag{7.9}$$

In Eq. 7.9 we used the fact that $\langle s_j \rangle = m$ for all j, where m is the average magnetization per spin. In the mean field approximation, we ignore the deviations of H_i from $\langle H_i \rangle$, and assume that the field at i is $\langle H_i \rangle$, independent of the orientation of s_i. This assumption is clearly an approximation because if $s_i = 1$, its neighbors will be more likely to be up. This fluctuation is ignored in the mean field approximation.

The effect of the mean field approximation is to reduce the original N-spin problem to an effective one-spin problem. It is straightforward to write the partition function for one spin:

$$Z_1 = \sum_{s_1=\pm 1} e^{-\beta s_1 \langle H_1 \rangle} = 2 \cosh \beta (qJm + H), \tag{7.10}$$

where we have used the fact that the cosh function is even. The free energy per spin is

$$f = -\frac{1}{\beta} \ln Z_1 = -kT \ln \left(2 \cosh \beta (qJm + H)\right), \tag{7.11}$$

and the magnetization per spin is

$$m = -\frac{\partial f}{\partial H} = \tanh \beta (qJm + H). \tag{7.12}$$

Equation 7.12 is a self-consistent equation whose solution yields m. That is, the mean field that influences the value of m also depends upon the value of m. It is instructive to write a little program to find the numerical solutions of Eq. 7.12 using a simple trial-and-error method. (Introduce the reduced temperature $\tilde{T} = \beta qJ$ and reduced field $\tilde{H} = \beta H$.) To investigate the possibility of spontaneous magnetization, set $H = 0$. It is easy to verify[?,?] that non-zero solutions of Eq. 7.12 exist for $H = 0$ when $\beta qJ \geq 1$. We know that the slope of the function $\tanh(\beta qJm)$ varies monotonically from its initial value βqJ to zero, and the slope of the left-hand side of Eq. 7.12 is always unity. Hence a non-zero solution of Eq. 7.12 exists only if $\beta qJ \geq 1$ or $kT \leq qJ$, and the critical temperature T_c that separates the $m = 0$ solution from the $m \neq 0$ solution is given by

$$kT_c = Jq. \tag{7.13}$$

For $H = 0$, the magnetization is small near $T = T_c$, and we can expand the $\tanh(\beta qJm)$ term in Eq. 7.12 to obtain

$$m = \beta qJm - \frac{1}{3}(\beta qJm)^3 + \ldots \tag{7.14}$$

Equation 7.14 has two solutions, $m = 0$, and

$$m = \frac{3^{1/2}}{(\beta qJ)^{3/2}}(\beta qJ - 1)^{1/2}. \tag{7.15}$$

The $m = 0$ solution corresponds to the disordered paramagnetic state and the $m \neq 0$ solution to the ordered ferromagnetic state. It is easy to verify (see Exercise 7.5) that the $m = 0$ solution corresponds to high temperatures, $T > T_c$, and the $m \neq 0$ solution corresponds to low temperatures, $T \leq T_c$. If we set $kT_c = qJ$, we find that for T near T_c, the magnetization vanishes as

$$m \sim \left(\frac{T_c - T}{T_c}\right)^{1/2}. \tag{7.16}$$

The behavior of other physical properties near T_c also is of interest. The zero-field susceptibility (per spin) χ is given by

$$\chi = \frac{\partial m}{\partial h} = \frac{\beta(1 - \tanh^2 \beta qJm)}{1 - \beta qJ(1 - \tanh^2 \beta qJm)}. \tag{7.17}$$

For T near T_c, we find

$$\chi \sim \frac{1}{T - T_c}. \tag{7.18}$$

The result Eq. 7.18 for χ is known as the Curie-Weiss law. Note that for very high temperatures, this result goes over to the Curie law, $\chi \sim 1/T$, for noninteracting spins.

The magnetization at T_c as a function of H can be calculated by expanding Eq. 7.12 to third order in H with $\beta = \beta_c = 1/qJ$:

$$m = m + \beta_c H - \frac{1}{3}(m + \beta_c H)^3 + \ldots \tag{7.19}$$

For very small m and H, we can assume that $\beta_c H << m$ and Eq. 7.19 reduces to the form

$$0 = \beta_c H - \frac{1}{3}m^3 \tag{7.20}$$

or

$$m = (3\beta_c H)^{1/3}. \tag{7.21}$$

The energy per spin in the mean field approximation is simply

$$e = -\frac{1}{2}qJm^2, \tag{7.22}$$

which is the average value of the interaction energy divided by two to account for double counting. Because $m = 0$ for $T > T_c$, the energy and hence the heat capacity also vanish for all temperatures $T > T_c$. Below T_c, the energy can be found by substituting Eq. 7.12 for m into Eq. 7.22. The result is

$$e = -\frac{1}{2}\,qJ\,[\tanh\,(\beta(qJm + H))]^2. \tag{7.23}$$

The specific heat C can be calculated from Eq. 7.23 for $T < T_c$. As shown in Exercise 7.5, $C \longrightarrow 3k/2$ for $T \longrightarrow T_c$ from below, and hence mean field theory predicts a jump in the specific heat at $T = T_c$.

We next compare the predictions of mean field theory with experiment. It is remarkable that a theory this simple can predict a critical point and the behavior of the system near the critical point. But it should not be surprising that the theory is too simple. The fact that the mean field result Eq. 7.13 for the critical temperature T_c depends only on the number of nearest neighbors q and not on the spatial dimension is one of the inadequacies of the theory. The simple mean field theory even predicts a phase transition for a one-dimensional Ising model, a prediction that is qualitatively incorrect. The mean field prediction for T_c for a square lattice is $kT_{c,\mathrm{mf}}/J = 4$ in comparison to the exact result, $kT_c/J = 2/\ln(1 + \sqrt{2}) \approx 2.269$. Another limitation of the simple form of mean field theory is that it predicts the average energy to vanish above T_c, a result that is clearly incorrect. These limitations are removed in more sophisticated mean field theories that treat the interaction of the central spin and its nearest neighbors exactly. However, as we discuss in section 7.5, all mean field theories yield quantitatively incorrect results for the behavior of the various thermodynamic functions near the phase transition.

7.5 Critical Exponents

Near T_c, mean field theory predicts that various thermodynamic properties exhibit power law behavior:

$$m(T) \sim (T_c - T)^{\beta} \qquad (T < T_c) \tag{7.24}$$

$$\chi(T) \sim |T - T_c|^{-\gamma} \tag{7.25}$$

$$m(T = T_c) \sim H^{1/\delta}. \tag{7.26}$$

The quantities β (not to be confused with the inverse temperature), γ, and δ are examples of *critical exponents*. Their mean field values are $\beta = 1/2$, $\gamma = 1$,

Table 7.1 Comparison of the critical exponents for the two- and three-dimensional Ising model with mean field theory.

Quantity	Exponent	$d = 2$ (exact)	$d = 3$	Mean field
Specific heat	α	0 (logarithmic)	0.113	0 (jump)
Order parameter	β	$\frac{1}{8}$	0.324	$\frac{1}{2}$
Susceptibility	γ	$\frac{7}{4}$	1.238	1
$M \sim H^{-1/\delta}$	δ	15	4.82	3
Correlation length	υ	1	0.629(4)	$\frac{1}{2}$
$c(r)$ at $T = T_c$	η	$\frac{1}{4}$	0.031(5)	0

and $\delta = 3$, respectively. These mean field values of the critical exponents are incorrect, except for four and higher dimensions (see Table 7.1). Mean field theory also predicts a jump in the specific heat, whereas experiments predict a power law or logarithmic divergence, e.g., $C \sim \log(T - T_c)$.

Although all mean field theories give incorrect values for the critical exponents, they correctly predict that many physical quantities exhibit power law behavior near the critical point. In general, we will suppose that any thermodynamic quantity can be separated into a regular part which remains finite near T_c and a singular part that is divergent or has divergent derivatives. The singular part will be assumed to be proportional to a power of ϵ, where ϵ is defined as

$$\epsilon = \frac{T - T_c}{T_c}. \tag{7.27}$$

We also will assume that a critical point can be characterized by an *order parameter*. For the Ising model the order parameter is the average magnetization, which satisfies the condition that $m = 0$ for $T > T_c$ and $m \neq 0$ for $T \leq T_c$.

In addition to characterizing the behavior of various thermodynamic functions by critical exponents, we can gain additional important information about the phase transition from the behavior of the spin correlation function c_{ij} defined as[7]

$$c_{ij} = \langle s_i s_j \rangle - m^2. \tag{7.28}$$

We have assumed the system is translationally invariant so that $\langle s_i \rangle = \langle s_j \rangle = m$. From the definition Eq. 7.28, we see that $c_{ij} = 0$ when the spin at i is uncorrelated with the spin at j. Hence, roughly speaking,

$$\sum_{j=2}^{N} c_{1j} \approx \text{number of spins correlated with spin 1.} \tag{7.29}$$

We have chosen spin 1 only for convenience. In Exercise 7.6 we show that the susceptibility χ is related to c_{ij} by

$$\chi = \frac{1}{NkT} \sum_{i,j=1}^{N} \left(\langle s_i s_j \rangle - \langle s_i \rangle \langle s_j \rangle \right) \tag{7.30}$$

$$= \frac{1}{NkT} \sum_{i,j=1}^{N} c_{ij} = \frac{1}{kT} \sum_{j=2}^{N} c_{1j}, \tag{7.31}$$

where we have used the fact that all lattice sites are equivalent. From Eq. 7.31, we see that the divergence of χ near T_c must be related to the behavior of c_{ij}. That is, because the right-hand side of Eq. 7.31 is related to the number of correlated spins, this number must increase as $T \longrightarrow T_c$ (and $\epsilon \longrightarrow 0$).

The above reasoning suggests that the neighborhood of the critical point is characterized by long-range correlations, i.e., large regions where the spins are correlated. You will be able to observe these correlated regions visually when you simulate the Ising model. To characterize these long-range correlations, we express c_{ij} as

$$c(r) = \frac{e^{-r/\xi}}{r^{d-2+\eta}}, \tag{7.32}$$

where ξ is called the correlation length, and r is the distance between i and j. Because χ increases as $T \longrightarrow T_c$, we expect that the correlation length ξ will increase also. This increase in ξ is characterized by the critical exponent v defined as

$$\xi \sim \epsilon^{-v}. \tag{7.33}$$

At $T = T_c$, ξ is infinite and $c(r)$ decays as a power law characterized by the critical exponent η; i.e., $c(r) \sim 1/r^{d-2+\eta}$ (see Eq. 7.32). Mean field theory predicts that $v = 1/2$ and $\eta = 0$ in contrast to their exact values $v = 1$ and $\eta = 1/4$ for the $d = 2$ Ising model.

The six critical exponents α, β, γ, δ, v, and η are not independent, but obey relations known as scaling relations. For example, in Exercise 7.7 it is shown that $\gamma = v(2 - \eta)$. The other known scaling relations imply that only two of the six critical exponents are independent.

7.6 Renormalization Group

The most important prediction of mean field theory is that the behavior of a system near a critical point is characterized by long-range correlations. We can understand the scaling relations by assuming that the correlation length ξ is the only important length of the system near T_c. This assumption implies that the lattice spacing is irrelevant. Because we know that ξ is infinite at T_c, we conclude that the system has no characteristic length at $T = T_c$.

These qualitative considerations can be used to develop a quantitative method for calculating quantities such as the critical exponents. This method is known as the renormalization group. For simplicity, we will introduce the ideas of the renormalization group for the $d = 1$ Ising model, even though this system has a phase transition only at $T = 0$.

Consider the Ising model for $d = 1$ with periodic boundary conditions. Its energy can be written as ($H = 0$)

$$E = -J \sum_{i=1}^{N} s_i s_{i+1}, \tag{7.34}$$

and the corresponding partition function is

$$Z = \sum_{\{s\}} \exp\left(\sum_{i=1}^{N} K s_i s_{i+1}\right), \tag{7.35}$$

where we have introduced the dimensionless parameter $K = -\beta J$. The idea is to average over the degrees of freedom on short length scales, because they are irrelevant to the behavior of the system.? One way to do such an average is to group sites into cells and then sum over cells. We write Z in Eq. 7.35 as

$$Z(K,N) = \sum_{\{s\}} e^{K(s_1 s_2 + s_2 s_3)} e^{K(s_3 s_4 + s_4 s_5)} \ldots \tag{7.36}$$

The form of Eq. 7.36 suggests that we can sum over even spins, e.g., $s_2 = \pm 1, s_4 = \pm 1, \ldots$. The result of this summation can be written as

$$\begin{aligned} Z(K,N) = & \sum_{\text{odd spins}} \left(e^{K(s_1+s_3)} + e^{-K(s_1+s_3)}\right) \\ & \times \left(e^{K(s_3+s_5)} + e^{-K(s_3+s_5)}\right) \ldots \end{aligned} \tag{7.37}$$

The other important idea is to write the partially summed partition function in Eq. 7.37 in the same form with $N/2$ spins and, in general, a different interaction K'. If such a rescaling is possible, we can develop a recursion relation for K' in terms of K. We require that

$$e^{K(s_1+s_3)} + e^{-K(s_1+s_3)} = A(K)\, e^{K' s_1 s_3}, \tag{7.38}$$

where the function A does not depend on s_1 or s_3. If the relation Eq. 7.38 exists, we can write

$$\begin{aligned} Z(K,N) &= \sum_{s_1, s_3, \ldots} A(K)\, e^{K' s_1 s_3} A(K)\, e^{K' s_3 s_5} \ldots \\ &= [A(K)]^{N/2} Z(K', N/2). \end{aligned} \tag{7.39}$$

For a large system we know that $\ln Z$ must be proportional to N, i.e., $\ln Z = N f(K)$, where $f(K)$ depends on K and is independent of N. From Eq. 7.39 we can express $\ln Z$ as

$$\ln Z = N f(K) = \frac{N}{2} \ln A(K) + \frac{N}{2} f(K'), \tag{7.40}$$

or

$$f(K') = 2 f(K) - \ln A(K). \tag{7.41}$$

The form of $A(K)$ can be found from Eq. 7.38 by noting that this relation must hold for all values of s_1 and s_3. The cases $s_1 = s_3 = \pm 1$ yield

$$e^{2K} + e^{-2K} = A\, e^{K'}. \tag{7.42}$$

For $s_1 = -s_3 = \pm 1$, we have

$$2 = A e^{-K'}. \tag{7.43}$$

From Eq. 7.42 and Eq. 7.43 we obtain

$$K' = \frac{1}{2} \ln \cosh 2K, \tag{7.44}$$

and

$$A(K) = 2 \cosh^{\frac{1}{2}}(2K). \tag{7.45}$$

We substitute the form of $A(K)$ in Eq. 7.45 into Eq. 7.41 to find

$$f(K') = 2f(K) - \ln[2 \cosh^{\frac{1}{2}}(2K)]. \tag{7.46}$$

Note that to within a factor of $-kT$, $Nf(K)$ is the Helmholtz free energy.

Equations 7.44 and 7.46 form a set of iterative equations for $f(K)$. In Exercise 7.8 we discuss how to use these equations to find $f(K)$ for all K and hence obtain the thermodynamic properties of the system.

The renormalization group method can be applied in higher dimensions, but as might be expected, it cannot be implemented exactly. In particular, it is not possible to write the partially summed partition function in exactly the same form as the original partition function.

7.7 Computer Simulations

The dependence of the energy on the spin configuration (see Eq. 7.1) does not tell us how the system changes from one spin configuration to another, that is, the Ising model does not have an intrinsic dynamics. However, as we learned on page 97 in chapter 5, we can interpret the Monte Carlo method as a fictitious dynamics, and we will find it convenient to talk about the evolution of the "time."

7.7.1 Metropolis Algorithm

The most popular Monte Carlo algorithm for generating configurations in the canonical ensemble (fixed T, H, N) was discussed in chapter 5 and is known as the Metropolis algorithm. For completeness we summarize it here. For an initial configuration of spins, choose a spin at random, and flip it. If the change in the total energy ΔE as computed from Eq. 7.1 is less than or equal to zero, accept the trial flip. Otherwise compute the quantity $w = \exp(-\beta \Delta E)$ and generate a uniform random number r in the unit interval. If $r \leq w$, accept the trial flip; otherwise retain the old spin configuration. This procedure is repeated until a sufficient number of independent configurations is obtained. The Metropolis algorithm generates configurations with the correct Boltzmann probability.

It is convenient to define the time in terms of Monte Carlo steps per spin, i.e., N trial flips equals one Monte Carlo step. That is, in one Monte Carlo step each of the N spins is visited once on the average. It is reasonable to interpret the Metropolis algorithm as a time-dependent process, because the flipping of individual spins resembles the real dynamics of a magnet whose spins are coupled to the vibrations of the lattice.

The most common boundary condition for computing the energy is the periodic boundary condition discussed in chapter 5. An example of the use of periodic boundary conditions is illustrated in Exercise 7.1. We adopt periodic boundary conditions in the program, but spin systems allow other possibilities (cf. ref. **?**).

Some of the equilibrium quantities of interest are the average energy E, the average total magnetization M, the heat capacity C, and the magnetic susceptibility χ. We will omit the brackets $\langle\ldots\rangle$ whenever no confusion will arise. We can calculate C at constant magnetic field either from its definition $C = \partial E/\partial T$, or from its relation to the fluctuations of the total energy:

$$C = \frac{1}{kT^2}(\langle E^2\rangle - \langle E\rangle^2). \tag{7.47}$$

Note that C can be interpreted as a measure of the response of the system to the addition of a small amount of energy. The relation Eq. 7.47 is an example of the relation of a linear response function to the equilibrium thermal fluctuations in the system. Another example of a linear response function is the zero field isothermal susceptibility given by

$$\chi = \lim_{H \longrightarrow 0} \frac{\partial M}{\partial H} \tag{7.48}$$

and its relation to the fluctuations of the magnetization:

$$\chi = \frac{1}{kT}(\langle M^2\rangle - \langle M\rangle^2). \tag{7.49}$$

The averages $\langle M^2\rangle$ and $\langle M\rangle$ in (7.49) are in zero field.

How do we know if two configurations are statistically independent? If two configurations differ by only the flip of one spin, they obviously are strongly correlated. One way of determining the time intervals over which configurations are correlated is to compute the time-dependent autocorrelation functions $C_m(t)$ and $C_e(t)$ defined as

$$C_m(t) = \frac{\langle M(t)M(0)\rangle - \langle M\rangle^2}{\langle M^2\rangle - \langle M\rangle^2}, \tag{7.50}$$

and

$$C_e(t) = \frac{\langle E(t)E(0)\rangle - \langle E\rangle^2}{\langle E^2\rangle - \langle E\rangle^2}. \tag{7.51}$$

$M(t)$ and $E(t)$ are the values of the magnetization and total energy of the system at time t, the number of Monte Carlo steps per spin. At $t = 0$, C_m and C_e

are unity. For sufficiently large t, $M(t)$ and $M(0)$ will become uncorrelated, and $\langle M(t)M(0)\rangle \longrightarrow \langle M(t)\rangle\langle M(0)\rangle = \langle M\rangle^2$. Hence $C_m(t)$ (and $C_e(t)$) should vanish for $t \longrightarrow \infty$. In general, the correlation functions decay approximately exponentially with t, e.g., $C_e \sim e^{-t/\tau}$, where τ is the correlation time. To keep the program simple, we have computed the desired physical quantities after every Monte Carlo step. However, because configurations separated by times less than τ are statistically correlated, this procedure is inefficient.

As mentioned above, the critical temperature of a square lattice is $kT_c/J \approx 2.269$. It is convenient to choose units such that Boltzmann's constant $k = 1$ and $J = 1$, and hence $T_c = 2.269$. What is the corresponding dimensionless magnetic field?

In Figure 7.1 we show the time-dependence of $C_e(t)$ for a square lattice with linear dimension $L = 4$. The total number of spins is $N = L^2 = 16$. The simulation is done at $T = T_c$ using the Metropolis algorithm. We see that $C_e(t) \approx e^{-t/\tau}$ with $\tau \approx 3.4$.

7.7.2 Finite Size Scaling Analysis

Simulations of finite-size systems suffer from an important limitation; that is, a finite system cannot exhibit a true phase transition characterized by divergent physical quantities. Instead, ξ and S reach a finite maximum at $p = p_c(L)$. For example, a quantity such as χ cannot have a true divergence at $T = T_c$, but must reach a finite maximum. However, as you will find in Exercise 7.11, even relatively small systems can exhibit behavior that is reminiscent of a phase transition. Because we can only simulate finite-size systems, how can we estimate the critical exponents from a computer simulation?

If $\xi(T) << L$, which holds if T is far from T_c, the measured values of the various physical quantities will not be affected by the finite linear dimension L of the lattice. Hence we might expect that a quantity such as χ will exhibit power law behavior if ξ is much less than L. However, if ξ is comparable to L, ξ cannot

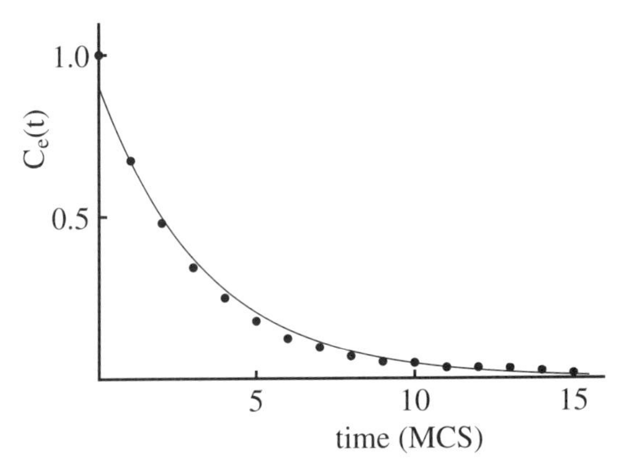

Figure 7.1: Time dependence of the energy autocorrelation function C_e for $L = 4$ using the Metropolis algorithm at $T = T_c$. The time is measured in Monte Carlo steps per spin. The run was for 10^4 Monte Carlo steps per spin.

continue to increase as $T \longrightarrow T_c$, and the power law behavior of χ will no longer be observed. This qualitative change in the behavior of χ will occur for

$$\xi(T) \sim L \sim \epsilon^{-v}. \tag{7.52}$$

We can invert Eq. 7.52 and write

$$\epsilon \sim L^{-1/v}. \tag{7.53}$$

Hence, if ξ and L are the same order of magnitude, we can replace Eq. 7.25 by the relation

$$\chi \sim L^{\gamma/v}. \tag{7.54}$$

We can use this dependence of χ on L at $T = T_c$ (Eq. 7.54) to determine γ by setting T to the critical temperature of the infinite lattice and determining χ as a function of L. If L is sufficiently large, we can use the asymptotic relation Eq. 7.54 to estimate the ratio γ/v. An example of such an analysis is shown in Figure 7.2 and is discussed in Exercise 7.12.

7.7.3 Demon Algorithm

The Metropolis or single-spin flip dynamics algorithm that we discussed in section 7.7.1 generates equilibrium configurations for a system in equilibrium with a heat bath at temperature T. Suppose that we instead wish to generate spin configurations at fixed total energy E (and fixed N and H). An obvious procedure is

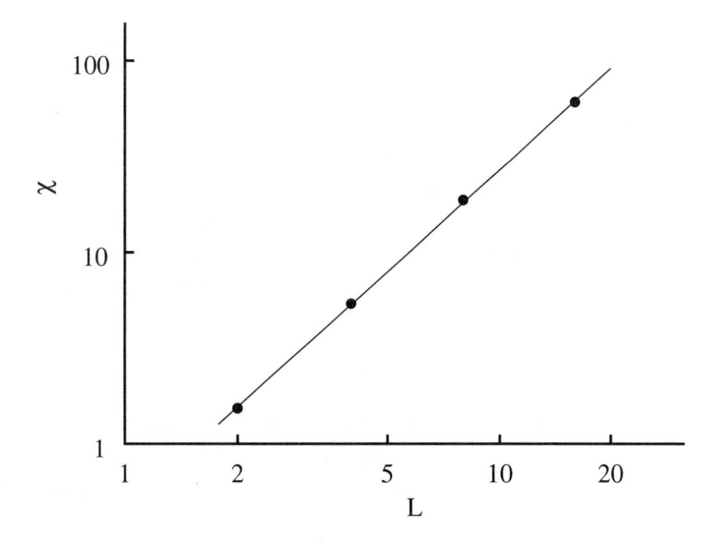

Figure 7.2: Log-log plot of the susceptibility (per spin) χ as a function of the linear dimension L for the Ising model on the square lattice at $T = T_c \approx 2.269$. Averages were performed over 10^4 Monte Carlo steps per spin. We find a slope of 1.77, which implies that $\gamma/v \approx 1.77$. If we use the exact result $v = 1$ for $d = 2$, we estimate $\gamma \approx 1.77$, an estimate that is consistent with the exact result $\gamma = 7/4$.

to begin with a configuration at the desired value of E, N, H and then randomly flip groups of spins until another configuration of the desired energy is reached. Such a procedure would be inefficient because we would have to reject most spin flips. An efficient Monte Carlo procedure for generating new configurations at fixed energy (the microcanonical ensemble) is due to Creutz. The idea is to imagine an extra degree of freedom, the demon, that can exchange energy with the system of interest. The only condition on the demon's energy is that it cannot be negative. The algorithm can be summarized as follows:

1. Begin with an initial spin configuration and demon energy E_d at the desired total energy.

2. Choose a spin at random and flip it.

3. Compute the change in the energy of the system, ΔE. If $\Delta E \leq 0$, the change is accepted and the energy is given to the demon, i.e., $E_d \longrightarrow E_d + |\Delta E|$. Return to step 2.

4. If the trial change increases the energy of the system, the trial configuration is accepted if the demon has sufficient energy to give to the system, i.e., $E_d \longrightarrow E_d - \Delta E$. Otherwise reject the trial change and retain the old configuration. Return to step 2.

The above procedure is repeated until a sufficient number of statistically independent configurations are obtained. Note that the total energy of the demon and the system remains a constant. Because the demon is only one degree of freedom in comparison to the many degrees of freedom of the system, we expect that the demon algorithm will yield representations configurations of the microcanonical ensemble in the $N \longrightarrow \infty$ limit. The demon algorithm offers a good opportunity to explore some basic questions about ensembles (see Exercises 7.14 and 7.15).

7.7.4 Cluster Dynamics

The existence of large correlated regions of spins near the critical point implies that the time required to flip these regions and obtain statistically independent configurations will increase with increasing ξ. This increase in the correlation time τ is called *critical slowing down* and is characterized by the critical exponent z defined as

$$\tau \sim \xi^z. \tag{7.55}$$

The Metropolis algorithm discussed in section 7.7.1 gives a value of $z \approx 2$. For a finite system, we expect that $\tau \sim L^z$, and hence the time required to find statistically independent configurations for large systems at $T = T_c$ becomes very long.

Physically, the reason for the large values of τ is that the Metropolis algorithm tries to break up large regions of correlated spins by flipping one spin at a time. The system would decorrelate faster if we could somehow flip more than one spin at a time. Recently, several *cluster acceleration* algorithms have been proposed that are much more efficient for large lattices and yield much lower values of z.

The simplest acceleration algorithm to implement is due to Wolff. In this algorithm a cluster of parallel spins is generated and flipped with probability unity. Spins are added to the cluster with probability $p = 1 - e^{-2J/kT}$. The algorithm, starting from an initial configuration, can be summarized as follows:

1. Choose a seed spin at random. The nearest-neighbor spins that are parallel to the seed spin are the perimeter spins.

2. Generate a random number r in the unit interval. A bond exists and a perimeter spin is added to the cluster if $r \leq p$.

3. If a spin is added to the cluster, find its perimeter spins that are not already part of the cluster.

4. Repeat steps 2 and 3 until no perimeter spins remain.

5. Flip all the spins in the single cluster.

The natural unit of time is the number of cluster flips.

The Wolff cluster algorithm yields a much smaller value of τ at $T = T_c$ for a given value of L. Hence, if your main interest is to obtain good estimates of various thermodynamic quantities in as short a time as possible, this algorithm is a good choice.

7.7.5 Spin Exchange Dynamics

The Ising model can describe many other systems that appear to have little in common with magnetism. For example, we can interpret the Ising model as a lattice gas, where "down" represents a lattice site occupied by an atom and "up" represents an empty site. Each site is occupied by at most one atom. Atoms interact only with their nearest neighbors. The Ising model also can be interpreted as a binary alloy whose two elements A and B may be substituted for one another on a set of fixed lattice sites. In this case the total number of A and B atoms is fixed and the Ising model exhibits an order-disorder transition.

The Metropolis and Wolff algorithms generate configurations at fixed T and the demon algorithm generates configurations at fixed E. In these three cases the total magnetization M is allowed to fluctuate. However, if we want to simulate a lattice gas or binary alloy, we need to conserve the number of down and up spins. The usual procedure is to use spin exchange or Kawasaki dynamics. In this algorithm a spin and one of its nearest neighbors are chosen at random and a trial interchange is made. The change in the energy ΔE is computed and the criterion for the acceptance or rejection of the trial interchange is the same as for the Metropolis algorithm.

7.8 Procedure for Running Program SPINS

The default screen and choices correspond to a Monte Carlo simulation using the Metropolis algorithm for a system of $N = 16 \times 16$ spins at the critical temperature of the infinite square lattice in zero external field. The initial configuration is generated at random and hence is not necessarily representative of a system at the critical point. The initial values of the total energy per spin and the magnetization per spin are shown in the right-bottom window. If the default choices are acceptable, the simulation can be started by either clicking the mouse in the lattice window or by pressing the hot key **F2.** The default is that up spins are shown in red and down spins are shown in green.

The total energy E is plotted as a function of the number of Monte Carlo steps per spin in the upper left window. Plots in yellow represent running averages and plots in green represent the value of the quantity at time t. The calculation of the running averages can be reset by pressing hot key **F4.** The plot can be switched between E and the magnetization M by clicking on the upper-left window.

The plot in the upper-right window is either the energy autocorrelation function C_e or the spin autocorrelation function C_m (see Eqs. 7.50 and 7.51). Because of the relatively small lattices that are used, the absolute value of M rather than M itself is used to determine C_m. The middle-right window shows either the probability distribution $P(E)$ that the system has energy E or the probability distribution $P(M)$ that the system has magnetization M. Both the correlation and probability plots can be switched between E and M by using the mouse to click on the respective windows. Clicking on these windows will also update the plots. The orientations of the spin are seen in the lower-left window. The lower-right window shows the number of Monte Carlo steps per spin (mcs), the average energy per spin E/N, the average magnetization per spin $m = M/N$, the specific heat (per spin) C determined from Eq. 7.47, and the susceptibility (per spin) χ determined from Eq. 7.49.

The menu headings allow the user to exit the program, change the simulation algorithm, and modify various settings. Under **File**, the user can read **About the Program** and **About CUPS**. A new configuration of spins can be chosen by selecting **New**, a previously saved configuration can be used by choosing **Open**, and an existing configuration can be saved in a file by selecting **Save**. One of the four algorithms can be chosen under the **Dynamics** menu heading. A demo can be seen at any time by choosing one of the four algorithms under the **Demo** heading. The demo mode can be exited by choosing an algorithm from the **Dynamics** menu. The configuration will be the same as the last configuration in the demo mode.

If the lattice is the proper size, that is, if $L_x = L_y = 2^p$, or 3^p, where the integer p is large enough to allow at least two renormalizations, the user can turn on **Renormalize** and see the original configuration and several renormalized configurations. A renormalized configuration is found by grouping the sites in the original configuration into cells or blocks of linear dimension b. The sign of the block spin is set by the majority of the spins in the block; ties are broken by a flip of the coin. More information about the meaning of the renormalized configurations is given in Exercise 7.13. The user can return to the default mode by selecting **off** from the **Renormalize** menu.

Under **Settings** the user can make changes in the **Parameters** (the temperature and the magnetic field), and in the **Display**. The temperature of the heat bath and the value of the magnetic field also can be set by sliders.

The function of the hot keys is similar to their role in Program MANYPART. Help can be called at any time by pressing hot key **F1**, and a run can be started or paused by pressing **F2**. Because the program will run faster if the spins are not shown on the screen, it is convenient to use hot key **F3** to toggle between showing and hiding this window. As stated above, **F4** resets the calculation of all averages. The orientation of individual spins can be changed by pressing hot key **F5** to enter the **Edit** mode. This mode can be used to obtain any desired spin configuration or energy. The user can return to the default mode by pressing hot key **F10**. Note that the menu is inactive in the edit mode.

The menu structure for the other algorithms is similar. Note that the temperature is not controlled by a slider in the demon algorithm, but the demon energy can be changed under **Settings**. The magnetic field is assumed to be zero in the Wolff cluster algorithm because its main utility is at the critical point. The magnetic field also is assumed to be zero in the Kawasaki spin exchange algorithm, but the total magnetization can be changed under **Settings**.

7.9 Exercises

Monte Carlo methods are very general and the Ising model is one of the most studied models in physics. The SPINS program can be applied in many ways that are not mentioned in the exercises or in the text (see references).

7.1 **Exact Enumeration of States for Small Systems**
Enumerate the 2^N microstates for the $N = 3$ and $N = 4$ Ising model in one dimension and find the corresponding contributions to Z_N for both free and periodic boundary conditions.

7.2 **Exact Solution of the One-Dimensional Ising Model**
Write Z_N for the $d = 1$ Ising model with free boundary conditions in the form

$$Z_N = \sum_{s_1=\pm 1} \cdots \sum_{s_N=\pm 1} e^{\beta J \sum_{i=1}^{N-1} s_i s_{i+1}} . \tag{7.56}$$

a. Use the form shown in Eq. 7.56 to show that $Z_2 = \sum_{s_1} \sum_{s_2} e^{\beta J s_1 s_2}$. The sum over s_2 can be done independently of s_1, and hence $Z_2 = \sum_{s_1} (e^{\beta J s_1} + e^{-\beta J s_1}) = \sum_{s_1} 2 \cosh \beta J s_1$. The sum over s_1 simply gives a factor of 2 and we find $Z_2 = 4 \cosh \beta J$ in agreement with Eq. 7.3.

b. Use the fact that spin N occurs only once in the exponential of Eq. 7.56 and show that

$$\sum_{s_N=\pm 1} e^{\beta J s_{N-1} s_N} = 2 \cosh \beta J . \tag{7.57}$$

Hence Z_N can be written as

$$Z_N = 2 \cosh \beta J \, Z_{N-1} . \tag{7.58}$$

Use the recursion relation Eq. 7.58 and the result Eq. 7.3 for Z_2 to verify the result (Eq. 7.5) for Z_N.

7.3 **Thermodynamic Properties of the One-Dimensional Ising Model**
Use Eq. 7.6 for F to confirm the following results for the entropy S, the internal energy E, and the heat capacity C of the one-dimensional Ising model:

$$S/k = N\left[\ln(e^{2\beta J} + 1) - \frac{2\beta J}{1 + e^{-2\beta J}}\right] \tag{7.59}$$

$$E = -NJ \tanh \beta J \tag{7.60}$$

$$C = Nk\,(\beta J)^2 \operatorname{sech}^2 \beta J\,. \tag{7.61}$$

What is the behavior of S in the limit $T \longrightarrow 0$ and $T \longrightarrow \infty$? What is the ground state of the system? What is the value of m in the ground state?

7.4 **Low-Temperature Behavior of the Heat Capacity**
Use the form of the heat capacity C found in Exercise 7.3 to plot the T-dependence of C. Why does C exhibit a maximum for $kT \sim J$?

7.5 **Mean Field Theory**

a. Substitute the two solutions for m into the expression Eq. 7.11 for the mean field free energy at $H = 0$ and expand the result in a power series. Verify that $m = 0$ provides a lower free energy for $T > T_c$, and that $m \neq 0$ provides a lower free energy for $T < T_c$.
b. Show that the specific heat C approaches $3k/2$ for $T \longrightarrow T_c$ from below.

7.6 **Relation of χ to Spin Correlation Function**
Derive the relation in Eq. 7.31 between χ and c_{ij}. Hint: Use the relation in Eq. 7.49 to show that χ can be expressed as

$$\chi = \frac{1}{kT}\langle (M - \langle M \rangle)^2 \rangle, \tag{7.62}$$

and write the total magnetization M as

$$M = \sum_{i=1}^{N} s_i\,. \tag{7.63}$$

7.7 **Scaling Relations**
Use Eq. 7.31 and the relations $\chi \sim \epsilon^{-\gamma}$ and $c(r) \sim \exp(-r/\xi)/r^{d-2+\eta}$ to show that $\gamma = \upsilon(2 - \eta)$. Hint: Convert the sum in Eq. 7.31 to an integral.

7.8 **Recursion Relations**
Equations 7.44 and 7.46 are a set of recursion relations that can be used to find $f(K)$ and hence the partition function for all K. If we know the partition function or $f(K)$ for one value of K, we can use Eq. 7.44 to find the new coupling constant K'. Because K' is less than K, we see that this way of using the recursion relations is equivalent to renormalizing the spin

configurations to higher and higher temperatures (recall that $K = J/kT$). We say that $K' = K = 0$ is a fixed point because further iterations of Eqs. 7.44 and 7.46 do not change K' and K. An alternative procedure is to start with the inverse of Eq. 7.44, i.e.,

$$K = \frac{1}{2}\cosh^{-1}(e^{2K'}). \qquad (7.64)$$

We also can solve Eq. 7.41 for $f(K)$ and use Eq. 7.43 for $A(K')$ to write

$$f(K) = \frac{1}{2}f(K') + \frac{1}{2}\ln 2 + \frac{1}{2}K'. \qquad (7.65)$$

Equations 7.64 and 7.65 are an equivalent set of renormalization group relations that allow us to start with a small value of K' (high temperature) and iterate to larger values of K until the fixed point $K = \infty$ $(T = 0)$ is reached. Suppose we start with $K' = 0.001$. For this small coupling constant, the interaction between the spins is negligible and hence we can assume that $Z_N(K'z = 0.001) = 2^N$ or $f(K' = 0.001) \approx \ln 2$. With this value of K' and $f(K')$, use Eqs. 7.64 and 7.65 to show that $K \approx 0.0316$ and $f(K) \approx 0.69365$. Use these values as the new K' and $f(K')$ quantities and use Eqs. 7.64 and 7.65 again to show that $K \approx 0.1797$ and $f(K) = 0.7092$. Continue iterating Eqs. 7.64 and 7.65 until $K \approx 2$. Make a table comparing your approximate results for $f(K)$ with the exact results given by Eq. 7.5.

7.9 **Approach to Equilibrium**

a. Choose the Metropolis algorithm with $L = 32$ and $H = 0$ and start from a random configuration. In such a configuration the orientation of each spin is chosen independently and the probability that a spin is up is 50%.). Let the system evolve at zero temperature. Estimate the time it takes for the system to reach equilibrium by looking at the $M(t)$ versus t plot.
b. Visually compare different equilibrium configurations at successively higher and lower temperatures. Are the configurations more-or-less ordered? If you start the system from a random configuration, from what temperature does such a start correspond?

7.10 **One-Dimensional Ising Model**
Use the Metropolis algorithm, $N = 32$, and $H = 0$ and find the dependence of the average energy E and average magnetization M on T in the range $T = 0.5$ to 5. What happens if you choose $H = 0.1$?

7.11 **Qualitative Behavior of the Two-Dimensional Ising Model**
Use the Metropolis algorithm to find the average values of E, M, C, and χ as a function of T (for $H = 0$) in the range $T = 1.5$ to 3.5 in steps of 0.25. Choose $L = 4$. Plot these quantities as functions of T and describe their qualitative behavior. Do you see any evidence of a phase transition? If time permits, repeat the simulation for $L = 8$.

7.12 **Finite Size Scaling**
Use the Metropolis or Wolff algorithms to determine C, $|m|$, and χ at

$T = T_c \approx 2.269$ for $L = 2, 4, 8$, and 16. Make a log-log plot of $|m|$ and χ to determine the critical exponents β and γ. Assume the exact result $\upsilon = 1$. Because the critical exponent $\alpha = 0$, C varies as $\ln L$. Is your data for C consistent with this result?

7.13 **Visual Renormalization Group**
Choose $L = 32$, $H = 0$, and $T = 3$ and either the Metropolis or Wolff algorithm. After the system has reached equilibrium, turn on the **Renormalize** option from the menu. How do the renormalized configurations compare? Then lower the temperature to $T = 2.0$ in steps of 0.2. Do you notice any qualitative change in the renormalized spin configurations?

7.14 **Demon Distribution Functions**

a. Given that there is no upper bound on the energy of the demon, why does the demon not get all the energy?
b. Run the simulation and confirm that the probability $P(E_d)$ that the demon has energy E_d is given by

$$P(E_d) \sim e^{-E_d/kT} \tag{7.66}$$

for all possible energies of the demon (see Exercise 7.1). (If necessary, use the **Settings** menu to change the upper limit of the demon energy that is plotted.) Do you find the result Eq. 7.66 for all lattice sizes?
c. How can you characterize the state of the demon? Explain the result Eq. 7.66 in terms of the nature of thermal equilibrium between a system and its heat bath.
d. What is the form of the probability distribution $P(M)$, where M is the magnetization of the system? Why is its form different than the form of $P(E_d)$?

7.15 **Average Demon Energy**
The form Eq. 7.66 of $P(E_d)$ implies that the average energy $\langle E_d \rangle$ of the demon is given by

$$\langle E_d \rangle = \frac{\sum E_d\, e^{-E_d/kT}}{\sum e^{-E_d/kT}}, \tag{7.67}$$

where the sums in Eqs. 7.67 are over the possible discrete values of E_d. Show that the possible energies of the demon are $0, 4J, 8J, \ldots$ and that the sums in Eq. 7.67 can be done exactly to yield

$$\langle kT/J \rangle = \frac{4}{\ln\left(1 + 4J/\langle E_d \rangle\right)}. \tag{7.68}$$

Why is $\langle E_d \rangle$ not proportional to kT as it is for a ideal classical gas?

7.16 **Ensembles**
The Metropolis or single-spin flip algorithm is claimed in the text to generate spin configurations in the canonical ensemble. However, if we interpret the Ising model as a binary alloy or lattice gas, the Metropolis algorithm would generate members of a grand canonical ensemble. Explain.

7.17 **Qualitative Estimate of the Critical Exponent z**
Quantitative estimates of z are made very difficult in part by the large lattices and run times that are required. These requirements are part of the reason that z is still not known with high confidence for the $d = 2$ Ising model, although we know that z is slightly greater than 2 for the Metropolis algorithm discussed in section 7.7.1. How do you expect the value of z for the demon algorithm and the Kawasaki spin exchange algorithm to compare to its value for the Metropolis algorithm?

7.10 Program Modifications

An artificial limitation of the program is that it is restricted to a square lattice of 48×48 spins. This restriction can be removed by changing the value of **Lmax**, if sufficient memory is available. The main physical quantity that is not computed in the program SPINS is the spin correlation function $c(r)$ defined in Eq. 7.28. It would also be of much interest to change the sign of J so that the antiferromagnetic Ising model could be studied and to consider a triangular lattice so that the phenomenon of frustration could be studied. Many other modifications could be suggested. However, if you want to go much beyond the capabilities of the program in its present form, you should write your own program.[5–7]

References

1. Reif, F. *Fundamentals of Statistical and Thermal Physics,* New York: McGraw-Hill, 1965.

2. Betts, D. S., Turner, R. E. *Introductory Statistical Mechanics,* Reading, MA: Addison-Wesley, 1992.

3. Kittel, C., Kroemer, H. *Thermal Physics,* 2nd ed. San Francisco: W. H. Freeman, 1980.

4. Chandler, D. *Introduction to Modern Statistical Mechanics,* New York: Oxford University Press, 1987.

5. Binder, K., Heermann, D. W. *The Monte Carlo Method in Statistical Physics,* New York: Springer-Verlag, 1988.

6. Gould, H., Tobochnik, J. *An Introduction to Computer Simulation Methods,* 2nd ed. Reading, MA: Addison-Wesley, 1995.

7. Heermann, D. W. *Computer Simulation Methods,* 2nd ed. New York: Springer-Verlag, 1990.

Appendix

Walk-Throughs for All Programs

These "walk-throughs" are intended to give you a quick overview of each program. Please see the Introduction for one-paragraph descriptions for all CUPS programs.

A.1 Walk-Through for FLUIDS

The program FLUIDS allows the user to explore the fluid properties of the van der Waals model and water. The user can choose four phase diagrams to display. The possible choices are v-T, p-T, p-v, u-T, u-v, s-T, and s-v diagrams, where v is the specific volume, p is the pressure, T is the temperature, u is the specific energy, and s is the specific entropy. The Helmholtz free energy $f(T, v)$ is used to calculate all the thermodynamic properties. For water an empirical formula for $f(T, v)$ is used. The primary goal is to draw a thermodynamic path in one phase diagram and see the corresponding path in other diagrams. In addition, the user can obtain thermodynamic quantities for any point on the phase diagram.

- After you start the program, data will be read in for the coexistence tables. These tables provide values of thermodynamic variables at the boundary of the two-phase coexistence region. When the tables are completed, the program credits along with a brief description of the program appears.

- After clicking the mouse or pressing any key, a screen will appear that allows you to specify water or the van der Waals model. On the same input screen, choose four phase diagrams to view simultaneously.

- Draw paths in the phase diagrams that are allowed, v-T for water, or any diagram except s-v and u-v for the van der Waals model. The program will then draw the corresponding paths in the other diagrams. To see an isothermal path, choose a diagram with temperature on the horizontal axis and draw a vertical line segment. An isobaric path can be drawn by drawing a vertical path in a phase diagram with pressure on the horizontal axis.

- Choose **F4-See Data** to collect thermodynamic data. The data that appears refers to the last point plotted. To collect data click the mouse at any point on a phase diagram, and then select the **F4-See Data** hot key.
- To clear the paths, choose **F2-Clear**.
- To change the type of fluid or the phase diagrams, choose the **F3-New** hot key.
- Choose **F10-Menu** to return to the main menu and exit the program.

The program FLUIDS is designed to allow users to explore the phase diagrams on their own. Some initial activities include drawing paths that cut across the two-phase coexistence region, drawing paths in only the gas- and only the liquid-phase region to see how their behavior in the other diagrams compare, and drawing a sequence of paths that form a closed thermodynamic cycle. By looking at the thermodynamic data from each of these cases, you can obtain a better feel for the typical magnitude of various thermodynamic quantities for the gas and liquid phases of matter.

A.2 Walk-Through for MANYPART Program

The program MANYPART is actually four programs in one. The user can choose between a molecular dynamics simulation of a system of particles at constant energy and density interacting via the Lennard-Jones potential, a molecular dynamics simulation of a system of hard disks at constant density, a Monte Carlo simulation of a system of particles at constant temperature and density interacting via the Lennard-Jones potential, and a Monte Carlo simulation of hard disks at constant density.

After the program credits are shown, the default screen is shown with a display of the initial positions of the $N = 16$ particles. Press hot key **F2** or click the mouse on the lower-left window to start the constant energy, molecular dynamics simulation.

- Let the simulation run for a while. The time dependence of the temperature and pressure is shown in the upper-right and upper-left window, respectively. Note that the temperature and pressure fluctuate in time (shown in green), but both quantities have well-defined averages (shown in yellow) if a sufficient number of configurations are included. Which quantity has larger relative fluctuations? Also shown is the probability density of the speed of the particles. What is the shape of the probability density of the speed?
- Click on the upper-left window to see a plot of the radial distribution function $g(r)$. Why is $g(r)$ near zero for $r < 1$? What is the interpretation of the different relative maxima of $g(r)$? (Remember that r is measured in terms of the length parameter σ of the Lennard-Jones interaction.) Click on the upper-right window to see a plot of the mean-square displacement $\langle R^2 \rangle$ (averaged over all particles). Does $\langle R^2 \rangle$ increase with the time t? Click on the middle-right window to see the velocity distribution. How does it differ from the speed distribution?

- Choose **Hard Disks** under the **Molecular Dynamics** menu. Why is it necessary to sometimes start from a lattice configuration? Run the simulation for a sufficient number of collisions to obtain a reasonable estimate for the mean pressure and mean free time.

- Choose **New** from the **File** menu and set the vertical and horizontal dimensions of the box equal to 4.5. What is the new density of the system? What is the maximum density for a hard disk system? How do you expect the pressure and mean free path to change? Compare your predictions to the results of the simulation.

- What is the meaning of the temperature in a molecular dynamics simulation? Why does it fluctuate if the particles interact with the Lennard-Jones potential and not fluctuate for hard disks? Return to the Lennard-Jones simulation and increase the temperature by 50 percent. Why is it not a good idea to increase the temperature too quickly?

- Suppose that you wish to simulate a system of particles interacting with the Lennard-Jones potential at a mean (dimensionless) temperature of $T = 1.0$. How easy is it to change the temperature to satisfy this condition?

- Choose **Lennard-Jones** from the **Monte Carlo** menu and set the mean temperature equal to 1.0 by either using the slider or by choosing **temperature** from the **Settings** menu. What does setting the temperature mean in the context of statistical mechanics? What is an advantage of a Monte Carlo simulation in comparison to a constant energy molecular dynamics simulation?

Quantitative results can be obtained if you let the program run sufficiently long. Special care must be taken to start the system and use should be made whenever possible of the initial conditions that are supplied with the program.

A.3 Walk-Through for QMGAS1 Program

The program QMGAS1 allows the user to calculate the thermodynamic properties and distribution functions for quantum ideal gases such as blackbody radiation, Debye model for phonons, ideal Fermi gas, ideal Bose gas, and the classical gas.

- After you start the program, the program credits along with a brief description of the program appears.

- After clicking the mouse or pressing any key, a screen appears that allows you to specify which system you wish to explore. The default values are for a three-dimensional Bose gas with the energy of the single particle states proportional to the square of the wave number.

- After accepting the input screen a window will appear asking for either a value of the fugacity z or the reduced temperature below the critical temperature T_c. If you are unsure what values to enter, press **Return** five times and random values will be entered for you. Then choose the fugacity button and press **Return** five more times. You will now have a table of thermodynamic data at ten different temperatures. Press **Cancel** to finish adding data.

- Now you can look at distribution plots for any three temperatures. Press **F2-E Plots**. When the input screen appears, press **Return**. You will see the plot of the density of states and plots of the various distribution functions. You can choose different temperatures by choosing **F2-E Plots** again and selecting the temperatures you want to see.

- Choose **F3-T Plots** to see plots of the chemical potential, energy, and the specific heat versus the temperature. Choose **F3-rescale** to improve the scales for the axes.

- You can add more data (**F5-Add Data**) to look more closely at a particular range of temperatures or you can choose **F10-Menu** to return to the menu and either exit the program or select another system to explore.

A.4 Walk-Through for QMGAS2 Program

The program QMGAS2 is a Monte Carlo simulation of a finite number of quantum ideal gas particles.

- After you start the program, the program credits along with a brief description of the program appear.

- After the mouse is clicked or any key is pressed, a screen appears that allows the user to specify the desired system. The default values correspond to a two-dimensional Bose gas of 50 particles with the energy of the single-particle states proportional to the square of the wavenumber.

- Select **F6-RUN** to start the simulation. You will see the positions in the state occupancy diagram change with time, indicating that particles are changing their states. You also will see in the lower right-hand corner a plot of the instantaneous energy per particle and the time averaged energy per particle, $< E >$. To see numerical results for various thermodynamic quantities and the distribution function $f(E)$, choose **F3-UPDATE**. Wait a few more seconds and choose **F3-UPDATE** again. You will see that the data has changed.

- Press **F5-SLOWER** a number of times to slow down the simulation. Then choose **F4-FASTER** to speed up the simulation.

- Select **F2-RESET** to cause the program to restart collecting data, but not change the state occupancy. This hot key is used when you think the system has reached equilibrium and you wish to collect useful equilibrium data.

- To stop the simulation temporarily, choose **F6-PAUSE**. To restart the simulation, choose **F6-RUN**.

- Use the slider to increase the temperature to about $T/T0 = 5$. This change will automatically reset the accumulation of data. Then decrease the temperature to $T/T0 = 0.1$. Note how the state occupancy changes.

- Either collect more data at different temperatures or explore another system. Choose **Specify System** from the **System** column of the menu to change the system to Fermi-Dirac statistics and repeat the above steps.

- Choose **File** and select **Exit Program** to exit the program.

A.5 Walk-Through for ISING Program

The program ISING incorporates four different algorithms for generating equilibrium configurations of the Ising model. The user can choose between the Metropolis, single-spin flip algorithm at constant temperature and constant external magnetic field; the Wolff, single-cluster flip algorithm at constant temperature (and zero magnetic field); the demon algorithm at constant energy and constant external magnetic field; and the Kawasaki, spin exchange algorithm at constant temperature and constant magnetization.

After the program credits are shown, the default screen is shown with a display of the initial orientations of the $N = 16 \times 16 = 256$ spins. Up spins are shown in red and down spins are shown in green. Press hot key **F2** or click the mouse on the lower-left window to start the simulation using the default Metropolis algorithm at the critical temperature of the infinite lattice and zero magnetic field.

- Let the simulation run for a while. The time dependence of the total energy is shown in the upper-right window. Note that the temperature (shown in green) fluctuates in time, but has a well-defined average (shown in yellow) if a sufficient number of configurations are included. The middle-right window shows a histogram of the energy of the system and the bottom right window shows the number of Monte Carlo steps (mcs), the mean energy and magnetization, the heat capacity, and the zero field susceptibility. Ignore the upper-right window for now.

- Click on the upper-left window to see the time dependence of the magnetization rather than the energy. Click on the middle-right window to see a histogram of the magnetization rather than the energy.

- Use the slider to lower the temperature to 0.1. What happens to the spins? Then raise the temperature and describe what happens. Make similar changes in the magnetic field and describe and explain what you see.

- Choose **Metropolis** under the **Demo** menu. Click on a spin to see the probability of the spin's being flipped. Compute this probability by doing a hand calculation and compare your results with those shown in the right-hand window. You also might wish to see a demo of the other algorithms to obtain a visual understanding of how these algorithms work.

- Return to the default screen (choose **Metropolis** under the **Dynamics** menu) and run the simulation for a while at the critical temperature of the infinite lattice ($T = 2.269$) and in zero magnetic field. Then choose **On** from the **Renormalize** menu. You will see the original lattice and several renormalized lattices. How do the lattices compare?

Quantitative results can be obtained if you let the program run sufficiently long. If you are interested in properties at the critical point, use the Wolff algorithm to obtain statistically independent configurations as quickly as possible.

A.6 Walk-Through for DIESEL, OTTO, and WANKEL Programs

Programs **DIESEL, OTTO** and **WANKEL** demonstrate the relationships between the movements of idealized physical engines and their thermodynamic properties. The engine's initial conditions are $T = 300$ K and $P = 1$ atm.

- The first screen initializes the inputs:
    1. The maximum operating temperature of the engine.
    2. The compression ratio.
- Select the default values: A maximum operating temperature of 1280.0 and a compression ratio of 15.5 for DIESEL and 9.5 for OTTO and WANKEL.
- **Press Enter or click on OK to run the program**
- The next screen is divided into three sections. The temperature versus entropy is plotted in the upper left section. The pressure versus volume is plotted in the upper right section. An animation of an engine cylinder's cycle and a description of the cycle are drawn in the lower section.
- **Select F1-Help hot key;** a help screen about the engine's cycle appears.
- **Select the F3-Step hot key (by mouse or keyboard)** to step through the engine's cycle.
- **Select the F5-Slower hot key (by mouse or keyboard)** to slow the engine.
- **Select the F6-Faster hot key** to speed the engine.
- **Select F10-Menu hot key** to access the commands in the menu or move the mouse cursor to the appropriate menu item.
- **Select Restart Program in the File drop down menu,** to return to the input screen or **Select Exit Program in the File drop-down menu** to exit program.

4.7 Walk-Through for ENGINE Program

ENGINE lets the user design an engine cycle by specifying the processes in the cycle, the engine type, and the gas type.

The first screen initializes the inputs:

1. The type of engine (reversible or irreversible).
2. The type of ideal gas used in the engine (helium, argon, nitrogen, or steam).
3. The initial temperature.
4. The initial pressure.
5. The percentage of heat loss (for an irreversible engine).
6. The output file name.

Nitrogen's properties are most like those of air. Room temperature is approximately 300 K.

- Use the default values and **press Enter or click on OK.**
- A new screen appears which is divided into four parts. In the upper left is the temperature (T) versus entropy (S) plot. In the upper right is the pressure (P) versus volume (V) plot. In the lower left is a summary table and in the lower right are four radio buttons labeled adiabatic, isobaric, isochoric, and isothermal.

  The Summary table lists the process (Proc, i.e., adiabatic, isobaric, isochoric or isothermal), the final pressure (P_f), the final volume (V_f), the final temperature (T_f), the final entropy (S_f), the work done by the engine, the heat absorbed by the engine, and the change in internal energy (ΔU) for each step in the cycle.

  Let's create an engine using the Diesel cycle.

  — **Select adiabatic**. The temperature, volume, and pressure sliders appear under the radio buttons. Use the temperature slider (or input the final value in the box to the right of the slider) to raise the temperature to around 800 K. Watch the T versus S and P versus V curves move as the slider is changed.

  — Then **select isobaric and continue to raise the temperature to around 1200 K.** Notice that the pressure slider disappears.

  — **Select the adiabatic again and return the volume back to its initial condition by moving the volume slider to its upper limit.** Notice the pressure slider appears again.

— Finally, **select isochoric and return either the pressure or the temperature to its initial conditions**. This closes the cycle. The radio buttons and sliders disappear and the **Engine Performance Summary** appears in its place.

- **Select F1-Help hot key;** a help screen about designing an engine cycle appears.
- **Select F10-Menu hot key** to access the commands in the menu or move the mouse cursor to the appropriate menu item.
- **Select Tutorial in the menu;** a drop-down menu appears. Click on the appropriate process for a description of that process. For example, click on **Adiabatic Proc.**
- **Select Restart Program in the File drop-down menu** to return to the input screen or **select Exit Program in the File drop-down menu** to exit program.
- After exiting program, open file ENGINE.DAT, and review the output.

A.8 Walk-Through for GALTON Program

A Galton board is a long, narrow board with an opening at the center of the top, a grid of pins, traps and/or reflecting walls just below the opening, and a series of bins at the bottom. Balls are dropped into the board from the opening at the top and are deflected by pins and/or reflecting walls or absorbed by the traps and/or absorbing walls. The final distribution of the balls in the bins is a function of the deflection probability (the probability that the ball will go right) of each pin and the positions of the traps, reflecting walls, and absorbing walls.

If the deflection probability is the same for all pins and there are no traps, reflecting walls and absorbing walls, the final distribution of the balls in the bins is described by a binomial distribution. As the number of balls becomes large, the binomial distribution approaches a normal distribution. In the program, this pin configuration is referred to as a traditional Galton board.

If the probability that a ball goes right or left depends on the pin and/or if there are traps, absorbing walls, and/or reflecting walls, the final distribution of balls is determined by using the laws of probability to compute the probabilities for every possible path to each bin. In the program, this pin configuration is referred to as a custom Galton board.

The central limit theorem states that the distribution of the means of a random sample approaches a normal distribution as the sample size becomes large. Applied to the Galton board, the central limit theorem states that, independent of the theoretical final distribution of balls in the bins, the distribution of the averages computed from a large number of similar Galton boards approaches the normal distribution. If the central limit theorem option is selected, the program tests the central limit theorem by repeating the previous Galton board 250 times, plotting the distribution of averages, and computing the chi-square between the distribution of averages and a normal distribution.

The first screen initializes the inputs:

1. The type of Galton board (traditional or custom).

2. The number of levels.

3. The number of balls which are dropped.

4. The deflection probability to the right (for a traditional Galton board).

5. The output file name.

- Select the default values: a traditional Galton board, six levels, 100 balls dropped, 50 deflections to the right, and an output data file of GALTON.DAT. **Press ENTER or click on OK to continue.**

  — Once the inputs have been defined, a screen appears with the Galton board on the left and statistical data on the right.

  For a small Galton board, one with 6 or fewer levels and 100 or fewer balls, the pins and the bins are shown (this case). For a large Galton board, only the bins appear.

  The statistical data include the predicted and actual final distribution of balls in the bins and the predicted and actual average of the distribution of balls in the bins. For a traditional Galton board, the statistical data also include the predicted and actual one sigma of the distribution of balls in the bins and, for a custom Galton board, the number of balls dropped and the predicted and actual number of balls in the bins.

  — **Select the F3-Step hot key (by mouse or keyboard);** for a small Galton Board, the balls slow down so you can observe their paths.

  — **Select F1-Help hot key;** a Help Screen appears.

  — **Select F10-Menu hot key** to access the commands in the **File** drop-down menu.

  — **Select Restart Program in the File drop-down menu** to return to the input screen.

- **Select a custom Galton board with 3 levels and 80 balls.**

  — This time a second input screen appears.
  The user specifies each pin's deflection probability to the right, and the positions of the traps, reflecting walls, and/or absorbing walls.

  — Place a **–1.** in the third row's middle pin. This puts a trap there.
  Before continuing, let's use the laws of probability to compute the predicted distribution of balls in the bins and the trap:

  All of the balls reach the top pin, where half the balls (50 percent of the balls dropped) should go to the left and half should go to the right.

  Half of the balls that reach the left pin in the second row (25 percent of the total balls dropped) should go to the left and half should go to the right, where they are absorbed by the trap.

Similarly, half of the balls that reach the right pin in the second row should go to the left, where they will be absorbed by the trap, and half should go to the right.

Half of the balls that reach the left-most pin in the third row (12.5 percent of the balls dropped) should fall into bin 0 and half into bin 1.

Half of the balls that reach the right most pin in the third row (12.5 percent of the balls dropped) should fall into bin 2 and half into bin 3. If 80 balls are dropped, 10 balls (or 12.5 percent) should be in bin 0, 10 balls in bin 1, 10 balls in bin 2, and 10 balls in bin 3.

The trap absorbs half of the balls dropped.

— **Press Enter or click OK** to run the program.

— Compare these predictions with the program's predictions and the actual results.

- **Select Cntrl Limit Thrm in the File drop-down menu,** to compute the chi square.
  - A new screen appears in which the distributions of the averages of 250 similar Galton boards are plotted on the left side of the screen.
  - The number at the top of the plot indicates the number of similar Galton boards which have been processed.
  - After all the Galton boards have been processed, the actual and theoretical average of the averages, the chi square, and the number of degrees of freedom appear on the right side of the screen.
  - Look in the chi square table in the text to determine the goodness of fit for the central limit theorem.

- **Exit Program (from the File drop down menu) and Open file GALTON. DAT.** Review the output.

A.9 The Kac Ring

The Kac ring was developed by Marc Kac to model a finite mechanical system that can be described by a statistical model with solvable equations of motion, time reversibility, and a Poincaré cycle. It provides a comparison between a physical system and its statistical description.

A Kac ring consists of a ring with a number, n, of evenly spaced points around the perimeter. At m of these points are markers. Between each pair of points is a ball that can be either dark or light. The equation of motion is, that as the balls move around the ring, they change color when they pass a point with a marker. It is a deterministic model that demonstrates time reversal and has a Poincaré cycle. Yet for large rings ($n >> 1$), there is good agreement between the Kac ring and its statistical model.

The canonical ensemble of the Kac ring is an ensemble of rings. Each ring has the same number of points and markers but a unique distribution of markers.

For a Kac ring, the quantity of interest is the number of dark balls $N(D)$ minus the number of light balls $N(L)$. The system is in equilibrium when $N(D) - N(L) = 0, 1$, or -1. For each turn of the ring (i.e., when a ball passes an adjacent point), $N(D) - N(L)$ is calculated exactly and compared to the canonical ensemble average $(\langle N(D) - N(L)\rangle)$. There are two options:

- A small Kac ring.
- A large Kac ring.
  The small Kac ring has nine balls and points. This option demonstrates that the Kac ring has a Poincaré cycle and time reversal. The large Kac ring has 2001 balls and points. It demonstrates that, as the number of balls and points becomes large, there is good agreement between the actual and statistical model.

A.9.1 Walk-Through for KAC Program

- The first screen initializes the inputs:
  1. The ring size.
  2. The number of markers.
  3. The initial number of dark balls.
  4. The number of moves of the ring.
  5. Output file name.
- Select the default values: a small ring, 2 markers, initially 2 dark balls, 18 moves, and output file name of KAC.DAT.
- **Press Enter or click OK** to run the program.
- The next screen is divided into two sections. The left section displays a small Kac ring and the right section displays a plot of $N(D) - N(W)$ and $(\langle N(D) - N(L)\rangle)$ versus the number of moves. The ring turns counterclockwise for 18 moves (a Poincaré cycle), then turns clockwise for 18 moves (simulating time reversal).
- **Select F1-Help hot key**; a help screen appears.
- **Select the F3-Step hot key (by mouse or keyboard)** to step through the ring's turns.
- **Select the F5-Slower hot key (by mouse or keyboard)** to slow the Kac ring.
- **Select the F6-Faster hot key** to speed the Kac ring.
  The purpose of this exercise is to demonstrate that the Kac ring has a finite Poincaré cycle (18 turns or twice the number of points in the ring) and time-reversal features. There, however, is not good agreement between $N(D) - N(W)$ and $(\langle N(D) - N(L)\rangle)$.
- **Select F10-Menu hot key** to access the commands in the **File** drop-down menu.

- **Select Restart in the File drop-down menu.**
The input screen appears.

- Selection a large Kac ring, 539 markers or any number or markers between 500 and 1500, initially 2 dark balls, and 100 turns.

- **Press Enter or click OK** to run the program.
A plot of $N(D) - N(W)$ and $(\langle N(D) - N(L)\rangle)$ versus the number of moves for 2001 turns (half a Poincaré cycle) is displayed. This plot is used to calculate the actual average and one sigma (or standard deviation) of $N(D) - N(W)$. The blue lines indicate the one sigma and the green lines indicate the two sigma deviations of $N(D) - N(W)$. When any key is pressed, this plot is redrawn for the first 100 turns. The yellow line plots $(\langle N(D) - N(L)\rangle)$ and the red line plots $N(D) - N(W)$. Note the close agreement between the two plots.

- **Exit Program (from the File drop-down menu) and Open file KAC.DAT.**
Review the output.

A.10 The Poisson Distribution

POISEXP uses statistical models to describe the decay rate of a radioactive substance and study the Poisson and exponential distributions. POISEXP simulates an ensemble of similar samples of 750 radioactive atoms. The half-life ($t_{1/2}$) is known. The method used to determine if an atom has decayed within a specified time interval (the observation time t_{obs}) involves generating a random number r and comparing it to the probability of a decay ($t_{obs} \times \ln(2)/t_{1/2}$ for the Poisson distribution or $t_{obs}\ln(2)/(\text{number of observations})$ for the exponential distribution). If r is greater than the probability of a decay, the atom decays.

There are two options:

- An illustration of the Poisson distribution.

- An illustration of the exponential distribution.

For the Poisson distribution option, the distribution of the (number of decays)/(allotted observation time) is plotted as a function of the number of decays. This distribution is compared to the corresponding Poisson distribution. For the exponential distribution option, the number of radioactive atoms as a function of time is compared to the corresponding exponential distribution.

A.10.1 Walk-Through for POISEXP Program

- The first screen initializes the inputs:

  1. Select distribution type: Poisson or exponential.
  2. Half-life.

- For the Poisson distribution:

  1. Observation time.
  2. Number of samples.

- For the exponential distribution:

  1. Observation time.
  2. Number of evenly spaced observations.
  3. Output file name and the output disk file name.

- Select the default values: the Poisson distribution with a half-life of 0.1 year and 200 samples of a substance consisting of 750 radioactive atoms.

- **Press Enter or click OK** to run the program.
  The next screen is divided into two sections. The upper section displaces the status of the atoms in a sample: gray—original radioactive atom (parent atom), white—decayed atom (daughter atom). The lower section plots the number of samples versus the number of decays in a sample during the observation time and the corresponding Poisson distribution. A box in the upper right corner of the lower section indicates which sample is shown in the upper section.

- **Select the F3-Step hot key (by mouse or keyboard)** to step through the samples manually.

- **Select the F5-Slower hot key (by mouse or keyboard)** to slow the stepping through the samples.

- **Select the F6-Faster hot key (by mouse or keyboard)** to speed the stepping through the samples.

- **Select F1-Help hot key,** and a help screen appears.

- Compare the program's predictions with the actual results.

- **Select F10-Menu hot key** to access the commands in the **File** drop-down menu.

- **Select Exponential in the Restart drop-down menu,** and an input screen for the exponential distribution appears.

- Select the default values: half-life of 0.01 years, an observation time of 1 half-life, and 10 evenly spaced observations.

- **Press Enter or click OK** to run the program.
  The next screen is divided into two sections. The upper section displaces the status of the atoms in the sample: gray—original radioactive atom (parent atom), white—decayed atom (daughter atom). The lower section plots the number of radioactive atoms (parent atoms) remaining versus time and the corresponding exponential distribution.

- Compare the program's predictions with the actual results.

- **Exit Program (from the File drop-down menu) and Open file POISEXP. DAT.** Review the output.

A.11 Walk-Through for STADIUM Program

The STADIUM program illustrates a system whose equation of motion is solvable but which demonstrates chaotic behavior. The system consists of a ball confined to a stadium-shaped container. The ball makes elastic collisions with the stadium's wall. It is a good representation of a two-dimensional ideal gas atom. The system's equation of motion is simply that the angle of incidence equals the angle of reflection. The ball's speed remains constant. The stadium model is a conservative system.

Since the ball's speed is constant, its trajectory can be specified by giving the ball's position and direction of motion when it hits the stadium wall. The position is specified by its location on the stadium's border s, and the direction of motion is specified by the cosine of the angle of incidence p. The quantity s can vary from 0 to $2(\pi+$ the length of the straight part of the stadium). (s, p) are conjugate coordinates. The ball's path can be described by a series of numbered pairs of $(s(n), p(n))$ where n is the number of bounces.

- The first screen initializes the inputs:

  1. Select the model: one-ball or two-ball model.
  2. The number of bounces.
  3. The length of the straight portion of the stadium: a number between 0 and 5, where 0 forms a circle, and 5 forms a long, thin oval.
  4. The ball's starting location along the perimeter of the stadium.
  5. The angle the ball's initial trajectory makes with respect to the x-axis.
  6. Output file name.

- Select the default values: the one-ball model, 20 bounces, the length of the straight portion equal to 2, an initial angle of 35, and output file name of STADIUM.DAT.

- **Press Enter or click OK** to run the program.
  The next screen is divided into two sections. The left section displays the stadium's outline and the ball's trajectory. The ball bounces 20 times. Its trajectory is drawn in white. Then it reverses its momentum (the angle between the ball's trajectory and the stadium's tangent at the point the ball hits). The ball bounces 20 more times. Its trajectory drawn in red.
  The right section displays a plot of the Poincaré Section (the momentum versus position). The points associated with the first twenty bounces are white and the last twenty bounces are red.

- **Select F1-Help hot key,** a help screen appears.

- **Select the F3-Step hot key (by mouse or keyboard)** to step through the ball's bounces.
- **Select the F5-Slower hot key (by mouse or keyboard)** to slow the ball.
- **Select the F6-Faster hot key,** to speed the ball.
  Because of time reversal, the ball returns to its initial location. The purpose of this exercise is to demonstrate that the chaotic nature of the stadium model is not due to cumulative computer computational errors.
- **Select F10-Menu hot key** to access the commands in the **File** drop-down menu.
- **Select Restart in the File drop-down menu.**
  The input screen appears.
- Select the two-ball model, 20 bounces, the length of the straight portion equal to 2, and an initial angle of 35.
- **Press Enter or click OK** to run the program.
  The next screen is divided into two sections. The left section displays the stadium's outline and two balls' trajectories. The white line represents the trajectory of the ball defined by the initial conditions and the red ball, its nearest neighbor. The two balls' trajectories have the same initial angle with respect to the x-axis but are separated by 0.1 units along the perimeter of the stadium. The separation in phase space as a function of the number of bounces is plotted in the right screen section. The balls bounce 20 times. The left screen is erased, and the process is repeated three more times: for a nearest-neighbor separation of 0.01 units, 0.001 units, and 0.0001 units.
  The purpose of this exercise is to show the chaotic behavior of the stadium model.
- **Exit Program (from the File drop-down menu) and Open file STADIUM. DAT.** Review the output.

A.12 Walk-Through for TWOD Program

The program TWOD simulates a two-dimensional unbiased random walk in which the "drunk" takes a specified number of steps. The drunk can move either on a grid or freely on the plane. The walk is repeated a specified number of times and the distributions of the final x direction displacement, the final y direction displacement, and the final radial displacement are plotted. The final x and y displacement distributions are compared to a binomial distribution. The final radial displacement distribution for the random walk on a plane is compared to a Rayleigh distribution.

- The first screen initializes the inputs:
  1. The type of walk: random walk on a grid or a plane.
  2. The number of steps in each random walk.
  3. The number of walks.
  4. Output file name.

- Select the default values: a random walk on a grid, 10 steps in each walk, 200 random walks, and output file name of TWOD.DAT.

- **Press Enter or click OK** to run the program.
  The next screen is divided into four sections. The upper left section displays the random walk on the grid. In the lower left and right section, the number of random walks versus the final x displacement and the number of random walks versus the final y displacement and the corresponding binomial distributions are plotted. In the upper right section, the number of random walks versus the final radial displacement is plotted.

- **Select F1-Help hot key,** and a help screen appears.

- **Select the F3-Step hot key (by mouse or keyboard)** to step slowly through the walks.

- Compare the program's predictions with the actual results.

- **Select F10-Menu hot key** to access the commands in the **File** drop-down menu.

- **Select Restart in the File drop-down menu**.
  The input screen appears.

- Select a random walk on a plane.

- **Press Enter or click OK** to run the program.
  The next screen is the same as the screen for a random walk on a grid; expect that the corresponding Rayleigh distribution is also plotted in the upper right section.

- Compare the program's predictions with the actual results.

- **Exit Program (from the File drop-down menu) and Open file TWOD.DAT**.
  Review the output.

Index

Limited Use License Agreement

This is the John Wiley & Sons, Inc. (Wiley) limited use License Agreement, which governs your use of any Wiley proprietary software products (Licensed Program) and User Manual(s) delivered with it.

Your use of the Licensed Program indicates your acceptance of the terms and conditions of this Agreement. If you do not accept or agree with them, you must return the Licensed Program unused within 30 days of receipt or, if purchased, within 30 days, as evidenced by a copy of your receipt, in which case, the purchase price will be fully refunded.

License: Wiley hereby grants you, and you accept, a non-exclusive and non-transferrable license, to use the Licensed Program and User Manual(s) on the following terms and conditions:

a. The Licensed Program and User Manual(s) are for your personal use only.
b. You may use the Licensed Program on a single computer, or on its temporary replacement, or on a subsequent computer only.
c. You may modify the Licensed Program for your use only, but any such modifications void all warranties expressed or implied. In all respects, the modified programs will continue to be subject to the terms and conditions of this Agreement.
d. A backup copy or copies may be made only as provided by the User Manual(s), but all such backup copies are subject to the terms and conditions of this Agreement.
e. You may not use the Licensed Program on more than one computer system, make or distribute unauthorized copies of the Licensed Program or User Manual(s), create by decompilation or otherwise the source code of the Licensed Program or use, copy, modify, or transfer the Licensed Program, in whole or in part, or User Manual(s), except as expressly permitted by this Agreement.
   If you transfer possession of any copy or modification of the Licensed Program to any third party, your license is automatically terminated. Such termination shall be in addition to and not in lieu of any equitable, civil, or other remedies available to Wiley.

Term: This License Agreement is effective until terminated. You may terminate it at any time by destroying the Licensed Program and User Manual together with all copies made (with or without authorization).

This Agreement will also terminate upon the conditions discussed elsewhere in this Agreement, or if you fail to comply with any term or condition of this Agreement. Upon such termination, you agree to destroy the Licensed Program, User Manual(s), and any copies made (with or without authorization) of either.

Wiley's Rights: You acknowledge that the Licensed Program and User Manual(s) are the sole and exclusive property of Wiley. By accepting this Agreement, you do not become the owner of the Licensed Program or User Manual(s), but you do have the right to use them in accordance with the provisions of this Agreement. You agree to protect the Licensed Program and User Manual(s) from unauthorized use, reproduction or distribution.

Warranty: To the original licensee only, Wiley warrants that the diskettes on which the Licensed Program is furnished are free from defects in the materials and workmanship under normal use for a period of ninety (90) days from the date of purchase or receipt as evidenced by a copy of your receipt. If during the ninety day period, a defect in any diskette occurs, you may return it. Wiley will replace the defective diskette(s) without charge to you. Your sole and exclusive remedy in the event of a defect is expressly limited to replacement of the defective diskette(s) at no additional charge. This warranty does not apply to damage or defects due to improper use or negligence.

This limited warranty is in lieu of all other warranties, expressed or implied, including, without limitation, any warranties of merchantability or fitness for a particular purpose.

Except as specified above, the Licensed Program and User Manual(s) are furnished by Wiley on an "as is" basis and without warranty as to the performance or results you may obtain by using the Licensed Program and User Manual(s). The entire risk as to the results or performance, and the cost of all necessary servicing, repair, or correction of the Licensed Program and User Manual(s) is assumed by you.

In no event will Wiley be liable to you for any damages, including lost profits, lost savings, or other incidental or consequential damages arising out of the use or inability to use the Licensed Program or User Manual(s), even if Wiley or an authorized Wiley dealer has been advised of the possibility of such damages.

General: This Limited Warranty gives you specific legal rights. You may have others by operation of law which varies from state to state. If any of the provisions of this Agreement are invalid under any applicable statute or rule of law, they are to that extent deemed omitted.

This Agreement represents the entire agreement between us and supercedes any proposals or prior Agreements, oral or written, and any other communication between us relating to the subject matter of this Agreement.

This Agreement will be governed and construed as if wholly entered into and performed within the State of New York.

You acknowledge that you have read this Agreement, and agree to be bound by its terms and conditions.